T0140219

Recent Advances in Biological Network Analysis

Byung-Jun Yoon • Xiaoning Qian
Editors

Recent Advances in Biological Network Analysis

Comparative Network Analysis and Network Module Detection

Editors
Byung-Jun Yoon
Department of Electrical
and Computer Engineering
Texas A&M University
College Station, TX, USA

Xiaoning Qian
Department of Electrical
and Computer Engineering
Texas A&M University
College Station, TX, USA

ISBN 978-3-030-57175-7 ISBN 978-3-030-57173-3 (eBook)
https://doi.org/10.1007/978-3-030-57173-3

© Springer Nature Switzerland AG 2021
This work is subject to copyright. All rights are reserved by the Publisher, whether the whole or part of the material is concerned, specifically the rights of translation, reprinting, reuse of illustrations, recitation, broadcasting, reproduction on microfilms or in any other physical way, and transmission or information storage and retrieval, electronic adaptation, computer software, or by similar or dissimilar methodology now known or hereafter developed.
The use of general descriptive names, registered names, trademarks, service marks, etc. in this publication does not imply, even in the absence of a specific statement, that such names are exempt from the relevant protective laws and regulations and therefore free for general use.
The publisher, the authors and the editors are safe to assume that the advice and information in this book are believed to be true and accurate at the date of publication. Neither the publisher nor the authors or the editors give a warranty, expressed or implied, with respect to the material contained herein or for any errors or omissions that may have been made. The publisher remains neutral with regard to jurisdictional claims in published maps and institutional affiliations.

This Springer imprint is published by the registered company Springer Nature Switzerland AG
The registered company address is: Gewerbestrasse 11, 6330 Cham, Switzerland

Preface

The diverse cellular mechanisms that sustain the life of living organisms are carried out by numerous biomolecules, including deoxyribonucleic acids (DNAs), ribonucleic acids (RNAs), and proteins. During the past decades, significant research efforts have been made to sequence the genomes of various species and to search these genomes to track down genes that give rise to proteins and noncoding RNAs (ncRNAs). As a result, the catalog of known functional molecules in cells has experienced a rapid expansion. Without question, identifying the basic entities that constitute cells and participate in various biological mechanisms within them is of great importance. However, cells are not mere collections of isolated parts. Biological functions are carried out by collaborative efforts of a large number of cellular constituents, and the diverse characteristics of biological systems emerge as a result of complicated interactions among many molecules. As a consequence, the traditional reductionistic approach, which focuses on studying the characteristics of individual molecules and their limited interactions with other molecules, fails to provide a comprehensive picture of living cells. To better understand biological systems and their intrinsic complexities, it is essential to study the structure, organization, and dynamics of molecular networks that arise from the complicated interactions among molecules within the cell.

In recent years, several high-throughput techniques for measuring molecular interactions—in particular, protein–protein interactions—have been developed for deriving network- or system-level understanding of cellular mechanisms. For example, the two-hybrid screening and co-immunoprecipitation followed by mass spectrometry have enabled the systematic study of protein interactions on a genome scale. Since protein–protein interactions are fundamental to all biological processes, a comprehensive protein–protein interaction (PPI) network obtained by mapping the protein interactome (complete set of protein interactions) provides an invaluable framework for understanding the cell as an integrated system. Furthermore, literature mining using natural language processing (NLP) techniques has become increasingly popular for searching through the vast amount of scientific literature to collect known molecular interactions. Nowadays, there exist a number of public databases—such as BioGRID (Biological General Repository for Interaction

Datasets), DIP (Database of Interacting Proteins), HPRD (Human Protein Reference Database), and STRING—that provide access to large collections of molecular interactions.

Considering the rapidly growing number and size of molecular networks, an important question is how we can utilize the available network data to gain novel biological insights. If we look back into the recent history of molecular biology research, there is no doubt that comparative sequence analysis methods and gene prediction methods have played central roles in analyzing the huge amount of genome sequencing data. In fact, such computational methods have been shown to be very useful for predicting novel genes and studying the organization of genomes. Similarly, comparative network analysis and network module detection techniques can serve as valuable tools for studying molecular networks. Such techniques can provide effective means of identifying novel functional modules (e.g., signaling pathways or protein complexes) that are embedded in molecular networks and conserved across different species, and they can also lead to important insights into the organization of molecular networks. Furthermore, various network-based analysis techniques can open up innovative ways of integrative omics data analysis that can ultimately lead to fundamental system-level insights into biological systems that may not be possible solely based on traditional reductionistic approaches.

In this edited book, we aim to overview recent advances in the emerging field of computational network biology—especially, comparative network analysis, network module detection, as well as network-based integrative omics data analysis in diverse biomedical applications. Leading international researchers in computational network biology who have been pushing the field forward have contributed to this book, where they review the latest techniques in network alignment, network clustering, network module detection, network-based integrative omics data analysis, discuss the advantages and disadvantages of the respective techniques, and present the current challenges and open problems in the field.

The edited book is organized based on the research topics. Specifically, Chaps. 1–4 review diverse comparative network analysis problems and the latest methods and algorithms for addressing them. Chapters 5–7 describe diverse mathematical formulations and computational solutions for functional module identification in molecular networks. Finally, Chaps. 8 and 9 focus on the application of computational network biology approaches for network-based integrative omics data analysis in biomedicine.

The primary audience for this edited book includes graduate students, academicians, and researchers who are currently working in areas relevant to computational network biology or those who wish to learn more about this emerging field. Furthermore, data scientists whose work involves the analysis of graphs, networks, and other types of data with topological structure or relations will also benefit from this edited book.

Finally, the editors would like to thank all the contributing authors. Without their outstanding contributions based on their expertise and experience on various research issues in computational network biology, this book project would not have been possible. We are also grateful to the fantastic Springer staff members for their tremendous editorial help as well as their patience while we slowly progressed from our initial ideas towards the production and publication of this edited book. Especially, we owe thanks to Ms. Merry Stuber who initiated the discussion about this book project and followed up with every step of this project from its inception until publication, and Ms. Maria David, who has provided enormous help while we worked on this book, meticulously checking its every detail and making many helpful suggestions.

Last but not least, we would like to thank our families for their love, encouragement, support, and sacrifice to give us time to finish this book.

College Station, TX, USA

Byung-Jun Yoon
Xiaoning Qian

Contents

Contributors

Hongryul Ann Department of Computer Science and Engineering, Seoul National University, Seoul, South Korea

Rishi M. Desai Institute for Systems Biology, Seattle, WA, USA

Rasha Elhesha University of Florida, Gainesville, FL, USA

Cesim Erten Antalya Bilim University, Antalya, Turkey

Ruiquan Ge Hangzhou Dianzi University, Hangzhou, China

Wayne B. Hayes Department of Computer Science, University of California, Irvine, CA, USA

Benjamin Hur Department of Computer Science and Engineering, Seoul National University, Seoul, South Korea

Dabin Jeong Interdisciplinary Program in Bioinformatics, Seoul National University, Seoul, South Korea

Hyundoo Jeong Department of Mechatronics Engineering, Incheon National University, Incheon, South Korea

Tamer Kahveci University of Florida, Gainesville, FL, USA

Gwangmin Kim Department of Bio and Brain Engineering, KAIST, Daejeon, Republic of Korea

Sun Kim Department of Computer Science and Engineering, Seoul National University, Seoul, South Korea

Yoo-Ah Kim National Center of Biotechnology Information, National Library of Medicine, NIH, Bethesda, MD, USA

Mijin Kwon Department of Bio and Brain Engineering, KAIST, Daejeon, Republic of Korea

Doheon Lee Department of Bio and Brain Engineering, KAIST, Daejeon, Republic of Korea
Bio-Synergy Research Center, Daejeon, Republic of Korea

Sangseon Lee Department of Computer Science and Engineering, Seoul National University, Seoul, South Korea

William J. R. Longabaugh Institute for Systems Biology, Seattle, WA, USA

Ji Hwan Moon Department of Computer Science and Engineering, Seoul National University, Seoul, South Korea

Minwoo Pak Department of Computer Science and Engineering, Seoul National University, Seoul, South Korea

Aisharjya Sarkar University of Florida, Gainesville, FL, USA

Yijie Wang Indiana University Bloomington, Bloomington, IN, USA

Qing Wu Hangzhou Dianzi University, Hangzhou, China

Jinbo Xu Toyota Technological Institute at Chicago, Chicago, IL, USA

Soorin Yim Department of Bio and Brain Engineering, KAIST, Daejeon, Republic of Korea

Part I
Comparative Network Analysis

Chapter 1
Global Alignment of PPI Networks

Cesim Erten

Abstract Given multiple PPI networks from different species, the *global PPI network alignment* problem is that of providing a global mapping between the nodes of the networks or subnetworks within them. Functional orthology detection, protein function prediction or verification, detection of common orthologous pathways, and reconstruction of the evolutionary dynamics of various species are some of the notable application areas of the global PPI network alignment problem. We focus on describing the basics of the problem, providing various formal definitions in the form of combinatorial optimization functions together with their computational complexities, and the algorithmic pillars of the suggested approaches. We also describe the common metrics employed in evaluating and comparing different global PPI network alignment outputs. Finally, we provide a discussion of relatively less studied aspects of the problem that may suggest potential open problems in need of further research on the topic.

Keywords Systems biology · PPI networks · Global network alignment

1.1 Introduction

Several high-throughput techniques including the yeast two-hybrid system [23], co-immunoprecipitation-coupled mass spectrometry [2], and computational methods such as those based on genome-wide analysis of gene fusion, metabolic reconstruction, and gene co-expression [30] enable extraction of large-scale protein–protein interaction (PPI) networks of various species. Several problem formulations related to network topologies [33], module detections [9], and evolutionary patterns [36] have been proposed for the analysis of these networks. From a comparative inter-

C. Erten (✉)
Antalya Bilim University, Antalya, Turkey
e-mail: cesim.erten@antalya.edu.tr

© Springer Nature Switzerland AG 2021

B.-J. Yoon, X. Qian (eds.), *Recent Advances in Biological Network Analysis*,
https://doi.org/10.1007/978-3-030-57173-3_1

actomics perspective, network alignment problems constitute yet another important family of problem formulations for the analysis of PPI networks.

In general terms, given two or more PPI networks from different species, where for each network nodes represent the proteins and the edges represent the interactions between the proteins, the *network alignment* problem is to align the nodes of the networks or subnetworks within them. Functional orthology is an important application that serves as the main motivation to study the alignment problems as part of a comparative analysis of PPI networks; a successful alignment could provide a basis for deciding the proteins that have similar functions across species. Such information may further be used in predicting functions of proteins with unknown functions or in verifying those with known functions [18, 62], in detecting common orthologous pathways between species [38], or in reconstructing the evolutionary dynamics of various species [42]. Before the introduction of network alignment as a model, common methods to detect orthologous groups of proteins have been solely based on the measures of evolutionary relationships usually in the form of sequence similarities. HomoloGene and Inparanoid [52] are examples of such approaches. Network alignment algorithms on the contrary incorporate the interaction data as well as the evolutionary relationships represented possibly in the form of sequence data. Based on the assumption that the interactions among functionally orthologous proteins should be conserved across species, such an incorporation is usually achieved by aligning proteins so that both the sequence similarities of aligned proteins and the number of conserved interactions are large.

Two versions of this general alignment framework have been suggested previously. In *local network alignment*, the goal is to identify from the input PPI networks, subnetworks that closely match in terms of network topology and/or sequence similarities. Approaches proposed for this version of the problem include PathBLAST [39], NetworkBLAST [60], MaWISh [40], Graemlin [25], and the graph match-and-split algorithm of [47]. Typically many overlapping subnetworks from a single PPI network are provided as part of the local alignments; this gives rise to ambiguity, as a protein may be matched with many proteins from a target PPI network. In *global network alignment* on the contrary, the goal is to align the networks as a whole. Starting with IsoRank [62], several algorithms employing various definitions have been suggested. Depending on the exact global network alignment problem definition, the desired output is in the form of unambiguous one-to-one mappings between the proteins of different networks or clusters of several proteins. A natural expectation is the design of efficient algorithms providing accurate alignment scores close to optimum values of appropriate formulations of the problem.

Many surveys and reviews on the topic of PPI network alignment have been published previously [15, 21, 24, 32, 46]. This chapter specifically considers the global network alignment problem and focuses on describing the basics of the problem, providing various formal definitions in the form of combinatorial optimization functions together with their computational complexities, and the algorithmic pillars of the suggested approaches. In Sect. 1.2 we provide necessary graph theory-related preliminary definitions. For the remainder, we consider three commonly employed global PPI network alignment problem versions; that is, balanced one-to-one, constrained one-to-one, and many-to-many alignments, in Sects. 1.3 through 1.5, respectively. For each version, we provide formal computational problem definitions, discuss their computational complexities, and provide a review of the proposed algorithms. In Sect. 1.6 we discuss the metrics employed in evaluating and comparing different global PPI network alignments. Finally, in Sect. 1.7 we provide a discussion of relatively less studied aspects of the problem that may suggest potential open problems in need of further research on the topic.

1.2 Preliminary Definitions

A PPI network is represented with an undirected graph $G(V, E)$, where V denotes the set of nodes corresponding to proteins and E denotes the set of edges corresponding to the interactions between proteins. For a graph $G(V, E)$, let $deg_G(u)$ denote the degree of $u \in V$. Let $\Delta(G)$ denote the maximum degree of any node of G. A *complete* graph is one where there is an edge between every pair of nodes.

A *matching* in a graph is a set of edges that pairwise do not have any common endpoint. If the set of edges E is weighted, the weighted version *maximum weight matching* corresponds to a matching with maximum sum of edge weights.

An *independent set* of $G(V, E)$ is a subset V' of V such that no two vertices in V' are adjacent. A *maximum independent set (MIS)* of G is an independent set of maximum size. If the set of nodes V is weighted, then the weighted version *maximum weighted independent set (MWIS)* corresponds to an independent set with maximum sum of node weights.

A *k-partite* graph $G(V = V_1 \cup V_2 \cup \ldots \cup V_k, E)$ consists of k disjoint parts $V_1, \ldots, V_k$, such that for an edge $uv \in E$, $u \in V_i, v \in V_j$, for $i \neq j$. A bipartite graph is a special case, where the node set consists of two parts. A *complete k-partite* graph is a k-partite graph where there is an edge between every pair of nodes each belonging to a different partition.

Given a set of graphs corresponding to PPI networks, $G_1, G_2, \ldots G_k$, let the *similarity graph* $S = (V_1 \cup V_2 \cup \ldots \cup V_k, E_S)$ denote an edge-weighted k-partite graph with parts $V_1, V_2, \ldots, V_k$ and edge set E_S. For an edge $uv \in E_S$, the weight of the edge, $w(uv)$, is an appropriately defined similarity function on a pair of proteins u, v, denoting the degree of homology between u and v. In its most general form an input instance of the global PPI network alignment problem is then denoted as $\prec G_1, G_2, \ldots, G_k, S \succ$. We note that the problem versions considered in Sects. 1.3 and 1.4 focus on the restricted case where $k = 2$.

1.3 Balanced Global One-to-One Network Alignment

Although an explicit formal definition is often omitted, informally the common goal in most of the previous global one-to-one PPI network alignment approaches is to provide a matching of the proteins of a pair of input PPI networks so that the resulting topological similarity, measured usually in terms of the number of conserved interactions between the matched proteins, is large and each pair of protein mappings in the matching contain proteins with high sequence similarity [3, 13, 14, 16, 20, 34, 35, 37, 42, 44, 48–50, 58, 61–63, 66].

1.3.1 Problem Definition

Formally, given $\prec G_1, G_2, S \succ$, we define the *Balanced Global One-to-One Network Alignment (BGONA)* problem as follows:

Definition 1.1 Find a matching M of S maximizing:

$$\alpha \times T(M) + (1 - \alpha) \times H(M). \tag{1.1}$$

The constant $\alpha \in [0, 1]$ is a balancing parameter that adjusts the relative importance of the network-topological similarity $T(M)$ that is induced by the matching and the homological similarities of the matched proteins, denoted by $H(M)$.

Several alternative definitions of $T(M)$ are employed in the literature. These alternative definitions can be grouped into two broad categories: *matching-independent* and *matching-dependent* topological similarity definitions. In the former, the score contributed by a matched protein pair uv to the overall $T(M)$ is independent of M and is only determined by the topologies of G_1 and G_2, whereas for the latter the contribution of uv depends on what other pairs are matched in M. Although missing from the relevant literature, such a categorization is quite important, as the computational complexity gap between alternative problem versions usually depends on the category of the employed topological similarity definition.

Regarding the *matching-independent* network-topological similarity definitions, two subcategories can be distinguished: one where $T(M)$ is formulated in terms of the similarities of subnetworks consisting of local neighborhoods of the matched proteins and the other where appropriately chosen network-wise global properties are employed. Definitions based on the so-called graphlet degree signatures, which take into account the number of local neighborhood subgraphs isomorphic to each distinct (up to isomorphism) graph of certain size, constitute the common instances of the former subcategory [41, 42, 45]. Definitions of the latter subcategory can be found in [34, 35, 42]. In [42], a similarity measure based on eccentricity, a network-wise global property, is employed in defining $T(M)$, in addition to the local neighborhood-based similarities including, degree, graphlet degree, and clustering coefficient. In [34, 35], $T(M)$ is a function reflecting how similarly "important" a pair of nodes are, where importance of a node is based on a minimum-degree heuristic definition, applied on the whole network globally.

With regard to the *matching-dependent* definitions, in almost all instances, the topological similarity score of an alignment is computed by checking the direct neighbors of the matched proteins in the alignment. In most of the relevant studies employing definitions in this category, $T(M)$ is defined as the size of the set of *conserved edges* [3, 13, 14, 20, 44, 61–63]. This set is defined as $\{(ux, vy) \in E_1 \times E_2 | uv, xy \in M \vee uy, xv \in M\}$. In some studies this definition is further refined by introducing appropriate normalizations. The goal in such normalizations is to avoid matching subnetworks with different densities when there is an almost equally edge conserving alignment, which matches similar-density subnetworks. In [49], assuming $|V_1| \leq |V_2|$, $T(M)$ is normalized by the size of the subgraph of G_2 induced by the nodes incident on edges in M. In [58], the difference between $|E_1|$ and the size of the conserved edge set is further added to the normalization factor. The exception to employ only the direct neighbors within *matching-dependent* similarity definitions is provided in [44], where each pair ux, vy in the conserved edge set contributes the average of the graphlet degree similarities of either uv, xy or uy, xv—depending on which pair is involved in M.

Regarding $H(M)$, almost all BGONA studies define it in terms of the sequence similarities of the matched proteins as, $\sum_{\forall uv \in M} w(uv)$. There seems to be a prevalent agreement on the definition of the node pair homological similarity function w; employed functions involve BLAST bit-scores or E-values of the involved pair of proteins. An exception to this prevalence is found in [34], where w is defined to be a normalized *cluster similarity* value of all pairs of clusters the mapped node pair belongs to, where pairwise cluster similarity is a function of sequence similarities of involved proteins. The clusters are obtained via a preprocessing step that involves hierarchical clustering of the input networks.

1.3.2 Computational Complexity

For the special case of $\alpha = 0$, the problem becomes that of maximizing $H(M)$, independent of the specific BGONA version that is employed. Regardless of the employed definition of the node pair homological similarity function w, maximizing $H(M)$ involves maximizing the sum of the edge weights in the matching of the bipartite similarity graph S. Since finding the *maximum weight matching in bipartite graphs* is solvable in polynomial time, so is the BGONA problem for this setting.

On the contrary for all other settings of $\alpha > 0$, the computational complexity of the BGONA problem essentially depends on the definition of $T(M)$ that is employed. In the case of matching-independent definitions, each of the node pair network-topological similarity functions, be it a local neighborhood-based score such as degree, graphlet degree sequence, and clustering coefficient or a global network-wise score such as eccentricity, minimum-degree heuristic-based score, can be computed in polynomial time. Assigning the weight of each edge in the bipartite graph S to be a convex combination of the homological similarity function w and the employed network-topological similarity function, the problem again turns into that of finding the maximum weight matching in a bipartite graph. Thus all of the matching-independent definitions give rise to polynomial-time solutions.

For matching-dependent definitions of $T(M)$ based on the size of the set of conserved edges, considering the special case of $\alpha = 1$, the BGONA problem definition becomes equivalent to a graph theory problem, that of finding the *maximum common edge subgraph (MCES)* of a pair of graphs. MCES is a problem commonly employed in the matchings of 2D/3D chemical structures [51]. The MCES of two undirected graphs G_1, G_2 is a common subgraph (not necessarily induced) that contains the largest number of edges common to both G_1 and G_2. The NP-hardness of the MCES problem proposed by Garey and Johnson trivially implies that the BGONA problem is also NP-hard in this setting of α and definition of $T(M)$ [28]. However it should be noted this result does not provide sufficient intuition regarding the computational complexity of the BGONA problem in actual practical settings. First of all, it does not suggest any network density-related argument. This is a drawback as the employed PPI networks are usually quite sparse. Second, it does not present the simultaneous nature of the BGONA problem aiming the optimization of two possibly conflicting goals in terms of $T(M)$ and $H(M)$. Assuming $T(M)$ is defined in terms of the size of the conserved edge set, the following result shows that the problem in its simultaneous optimization nature is hard even when the input graphs G_1, G_2 are simply paths, thus handling both drawbacks [3]:

Theorem 1.1 *For any* $0 < \alpha < 1$, *the BGONA problem is NP-hard, even when* G_1, G_2 *are paths.*

Note that assigning $\alpha = 0$ or 1, the problem is trivially polynomial-time solvable for this setting, where $G_, G_2$ are paths. Although similar results for other versions of matching-dependent definitions of $T(M)$ are not known, since the theorem limits the inputs to simply paths, we conjecture that it can be extended to them via trivial reductions from this setting.

1.3.3 Algorithms

We provide a brief review of the algorithms proposed for each version of the BGONA problem. Those that employ a matching-independent definition based on graphlet degree sequences are GRAAL [41], MI-GRAAL [42], and H-GRAAL [45]. GRAAL first constructs a graphlet degree sequence for every node u in G_1 and v in G_2. This is done by counting the number of distinct subgraphs (up to isomorphism), called graphlets, that the node touches, for all 2- to 5-node graphlets. Similarity of each node pair u, v is computed from their graphlet degree sequences. Next a cost is assigned to each node pair u, v based on this similarity measure and $deg_{G_1}(u), deg_{G_2}(v)$ so that more "similar" node pairs both with large degrees provide a lower cost than those with small degrees. GRAAL then applies a greedy seed-and-extend heuristic to form the alignment; once the minimum cost pair is selected as the seed alignment, the current alignment is extended by considering the *candidate set*, which are the not-yet-aligned node pairs directly connected to nodes of G_1, G_2 in the current alignment and among them choosing the node pairs with minimum cost. The process is repeated on *2-extensions* G_1^2, G_2^2 and *3-extensions* G_1^3, G_2^3, where all nodes of distance less than or equal to i in a graph G are connected by a direct edge in the *i-extension* of G, denoted by G^i. MI-GRAAL is quite similar to GRAAL with two main differences. First, MI-GRAAL employs a combination of degree, clustering coefficient, graphlet degree sequence, eccentricity, and sequence similarity to define the similarity of a node pair. Second, although MI-GRAAL also uses a greedy seed-and-extend heuristic similar to GRAAL, different from GRAAL, MI-GRAAL employs a maximum weight bipartite matching algorithm on the candidate set to extend a current alignment. H-GRAAL uses the same cost computations as GRAAL. However rather than employing a greedy heuristic to form the alignment, it employs the Hungarian algorithm to find the optimum alignment minimizing total cost, which is the same as the minimum weight bipartite matching. To generalize the solution to finding all optimum alignments, a dynamic version of the minimum weight bipartite matching algorithm is used. In HubAlign [35], first an "importance" value is computed for each node, where importance indicates the network connectivity of a node. A minimum-degree heuristic method similar to the tree decompositions of graphs is suggested for this purpose. The topological similarity score of a node pair is assigned as a function of the similarity of the two nodes' importance values. The contribution of a node pair to the overall alignment score is then provided as a convex combination of this topological similarity and a normalized BLAST bitscore-based sequence similarity. The actual alignment is then constructed following a greedy seed-and-extend heuristic similar to that of GRAAL, except once an iteration of extending the alignments greedily are finished, rather than considering the i-extensions of the graphs, it simply selects a new *best* seed to repeat the procedure of extending the alignment. ModuleAlign [34] employs the same node pair topological similarity function and defines the alignment score as a convex combination of it and a hierarchical clustering-based homology score. The actual alignment construction is done in two steps. In the first step similar to H-GRAAL, the Hungarian algorithm for maximum weight bipartite matching is used to find an initial alignment. Next a heuristic is applied repeatedly to increase the edge conservation of the initial alignment. This is done via removing the current alignment of a neighbor u' of u in the maximum scoring pair uv, updating the alignment score of $u'v'$, for each neighbor v' of v, and re-optimizing the maximum weight bipartite matching with one step of the Hungarian algorithm.

Among the approaches based on the matching-dependent topological-similarity definitions of BGONA, many consider the conserved edges version. SPINAL [3] is one such approach, which consists of two phases: a coarse-grained alignment score estimations phase and a fine-grained conflict resolution and improvement phase. Both phases make use of the construction of neighborhood bipartite graphs and a set of contributors as a common primitive. Using these concepts within iterative local improvement heuristics constitutes the backbone of the algorithm. PISwap [14] employs a local search heuristic to improve an initial alignment obtained via the maximum weight bipartite matching of the similarity graph consisting of homological similarity-weighted edges with the Hungarian algorithm. The search heuristic repeatedly searches for a pair of node mappings $uv, u'v'$ so that swapping the mapping to $uv', u'v$ improves the overall alignment score. In Natalie [20], an integer linear programming (ILP) formulation of the BGONA problem, generalizing the quadratic assignment problem (QAP), is formulated. The main strategy for updating the Lagrangian multipliers combines both the subgradient optimization and the dual descent method. Similarly L-GRAAL [44] proposes an alignment search heuristic based on integer programming and Lagrangian relaxation. Such a heuristic is employed in extending seed alignments, which are computed based on a score consisting of a convex combination of node pair homological similarities and graphlet degree sequence-based node pair topological similarities. IsoRank [61–63] considers a random walk on the product graph $G^*(V^* = V_1 \times V_2, E^*)$, where there is an edge between nodes $(u, v) \in V^*$ and $(x, y) \in V^*$, if the pair induces a conserved edge, that is, $ux \in E_1$, $vy \in E_2$. As usual, the random walk moves from a node (u, v) in G^* to any of its neighbors with the same probability. Equivalently, the problem is posed as an eigenvalue problem as well. A solution based on the iterative power method to solve the corresponding eigenvalue problem is proposed [31].

1.4 Constrained Global One-to-One Network Alignment

For an instance $\prec G_1, G_2, S \succ$, where S is a possibly complete bipartite graph, denoting pairwise protein sequence similarities, the constrained global one-to-one pairwise network alignment setting assumes a preprocessing, which involves an appropriate filtration of S into S' so that for given integers k_1, k_2 we have $deg_{S'}(u) \leq k_1$ for $u \in V_1$ and $deg_{S'}(v) \leq k_2$ for $v \in V_2$. The preprocessing is designed so as to retain only large-weight edges in S'. Thus the need to consider specific edge weights in the similarity graph might be eliminated, which leads to a simplified optimization function. Furthermore the bipartite graph to consider now, that is S', shall be degree-bounded.

1.4.1 Problem Definition

Formally, given $\prec G_1, G_2, S' \succ$ such that for k_1, k_2 integer constants, $deg_{S'}(u) \leq k_1$ for $u \in V_1$ and $deg_{S'}(v) \leq k_2$ for $v \in V_2$, we define the *Constrained Global One-to-One Network Alignment (CGONA)* problem as follows:

Definition 1.2 Find a matching M of S' maximizing $T(M)$.

We note that in almost all CGONA approaches $T(M)$ is defined to be the size of the set of conserved edges. An exception is the version considered in [10], where to tolerate a certain amount of missing interaction data, "indirect" links up to a certain distance are also allowed; if a pair of proteins u, x interacts in one of the graphs, say G_1, and the other pair of proteins v, y is at most distance two in G_2, then the pair of matchings uv and xy is still considered to induce a conserved edge.

Regarding the preprocessing involving appropriate filtration of S to arrive at degree-constrained S', one alternative is that of [1]. The filtration is done via *maximum weight b-matching in bipartite graphs* (or the *degree-constrained subgraph problem*), which has been studied fairly well starting with the pioneering work of [19]. Polynomial time solutions, including appropriate modifications of the network flow algorithms [26] and belief propagation methods [11], have been suggested. On the contrary [10] and [66] employ the clustering produced by the Inparanoid algorithm [53], which produces non-overlapping clusters of orthologous sequences using pairwise BLAST sequence similarity scores. A node $u \in V_1$ is then allowed to be possibly mapped to one of V_2 nodes in the same cluster as v_1, and vice versa. In our setting the cluster with the most V_2 nodes determines k_1, whereas one with most V_1 nodes determines k_2. With this approach, there is no direct parameter that sets k_1, k_2 as in b-matching. Nevertheless the clusters produced by Inparanoid tend to be small in practice, giving rise to reasonable degree constraints k_1, k_2.

1.4.2 Computational Complexity

Although the CGONA version of the problem simplifies the optimization goal of Eq. 1.1 by eliminating the sequence similarities from the optimization score, the following result indicates that the problem remains computationally hard even in this setting [22]:

Theorem 1.2 *The CGONA problem is APX-complete even when* $k_1 = 2, k_2 = 1$, *and* G_1, G_2 *are bipartite graphs with* $\Delta(G_1) \leq 3$ *and* $\Delta(G_2) \leq 2$.

In spite of this negative computational complexity result applying to even quite restricted problem instances, for certain instances some positive theoretical results, especially in terms of fixed-parameter tractability, are still possible. In particular, the following results are provided in [4]:

Theorem 1.3 *The CGONA problem is fixed-parameter tractable when* k_1 *is any fixed positive integer constant and* $k_2 = 1$.

Theorem 1.4 *For any constant* k_1, k_2, *the CGONA problem is fixed-parameter tractable when* G_1, G_2 *are bounded-degree graphs.*

Even though valuable theoretically, since the fixed parameter is assumed to be the size of the output set, that is $T(M)$ maximizing the input instance, and the suggested approaches require exponential running times in terms of $T(M)$, such results are of practical value only for aligning PPI networks of distant species, where the maximum number of conservable edges is small.

1.4.3 Algorithms

Approaches considering the CGONA problem include [1, 4, 10, 22, 64, 66]. We note that in [10, 64, 66], the special case of *cluster-restriction*, which contains further restrictions on S', is considered. If $uv, uy, xv \in S'$, the restriction implies that $xy \in S'$ as well. Under this constraint, S' consists of a disjoint set of cliques, called *clusters*.

The notion of maximum independent sets (MIS) is a common algorithmic ingredient in the independent sets-based approaches of [1, 4, 22]. Let c_4 denote a specific 4-cycle $uxvy$, where $ux \in E_1$, $vy \in E_2$, and $uy, xv \in E_S$ are similarity edges. We say that two c_4s *conflict*, if at least two of their similarity edges are adjacent but distinct, which implies that they cannot coexist in any matching M of S. Given an instance $\prec G_1, G_2, S \succ$, we can construct its *conflict graph*, where the nodes correspond to the c_4s of the instance and there is an edge between a pair of nodes if their respective c_4s conflict. For example, for an instance $\prec G_1, G_2, S \succ$, where G_1 consists of the edges ab, bc, G_2 consists of the edge gh, and S consists of the edges ag, bg, bh, cg, ch, we have three c_4s, $abhg$, $bcgh$, and $bchg$. Since each pair of these c_4s conflict, the resulting conflict graph is a triangle. With such a construction, it is easy to see that maximizing $T(M)$ in the instance $\prec G_1, G_2, S \succ$ is equivalent to maximizing the size of the independent set of the resulting conflict graph. The model can further be extended by assigning a certain importance to each c_4 reflecting the values of the involved pair of matchings. The problem in this case becomes equivalent to the weighted generalization of the MIS, maximum weighted independent set (MWIS). Each independent sets-based approach then differs in the choice of the model, MIS or MWIS, and the appropriate methods employed to solve it. Detailed theoretical investigations of the problem under the MIS model are made in [4, 22]. In [4], several structural properties of conflict graphs are presented for various cases of k_1, k_2. For instance, it is shown that the conflict graphs are free of induced wheels of size greater than 6 and induced fans of size greater than 7, for $k_2 = 1$ and any constant k_1. Such results give rise to efficient polynomial-time algorithms to approximately optimize the MIS of the conflict graph, which leads to efficient approximation algorithms for the CGONA problem. On the contrary, CAMPways proposes a MWIS-based model [1]. Although in CAMPways the alignment of directed graphs in the form of metabolic pathways is considered, the overall methodology and the employed techniques are readily applicable to the alignment of undirected graphs as well. Since the constructed conflict graph is node-weighted, the MWIS model applies. Restricting the model to undirected graphs, the weight of a conflict graph node is the average sequence similarity of the matched node pairs of the corresponding c_4. Since MWIS is NP-complete, the GWMIN2 heuristic defined in [57] is used to find a large weight-independent set of the conflict graph, which corresponds to the output alignment produced by the algorithm.

A method based on Markov Random Field (MRF) is proposed in [10] for the cluster-restriction of CGONA. This model is specified by a $\prec G_1, G_2, S \succ$ and conditional probability distributions that relate the event that a given pair u, v is in the optimum matching with those events for the pair's neighbors in their respective PPI networks. A binary variable z_{uv} is associated with each $uv \in S$, which indicates the probability of the match being in the optimum alignment. The *neighbors* of uv are those pairs xy that induce a conservation with uv. Then the conditional probability of z_{ij}, given the probabilities of its neighbors, is defined as a function of α, β and the probabilities of the neighbors. The parameters α and β are estimated on the basis of training data, then a Gibbs sampling is performed to define the values of unknown variable z on the test set. Another method suggested for the cluster-restriction version of CGONA is proposed in [66]. It is based on a message passing approach, inspired by the Viterbi algorithm [65]. A *cluster graph* is defined as one where each node corresponds to a cluster c_i and there is an edge between nodes corresponding to c_i and c_j, if $ux \in E_1, vy \in E_2$, for some $u, v \in c_i$ and $x, y \in c_j$. The messaging passing is applied on this graph. Assuming the cluster graph is acyclic, assigning a node as the root defines a tree. Starting from the leaves of the tree, a forward sweep toward the root is done, where each cluster node computes the maximum number of conserved edges possible for every possible matching of the nodes it contains based on the numbers of conserved edges of its children. Once the maximum number of conserved edges for the root cluster node is determined, the optimal local matching of a cluster node is obtained by recovering the matching that achieved the optimal number of conserved edges for the optimal local matching of the parent cluster node through a backward step from the root toward the leaves. Such a message passing algorithm on an acyclic graph is polynomial time, assuming constant k_1, k_2. HopeMap constructs the same cluster graph and finds its connected components. Each component is scored via a combination of induced conserved edges, homology score, and functional coherence score defined in terms of Gene Ontology (GO) biological process terms covered by the nodes in a connected component.

1.5 Global Many-to-Many Network Alignment

A central informal objective in global many-to-many network alignments is the construction of clusters of orthologous proteins that guarantees that each cluster is composed of members with high homological similarity, determined via sequence similarities, and that the interactions of the proteins involved in a pair of clusters are conserved across the input networks. Such a problem version, as compared to the one-to-one version, is especially useful in cases of gene duplication across species.

Remark 1.1 Most of the relevant alignment studies and surveys classify the GMNA approaches under the *global multiple network alignment* category, where the emphasis is on aligning more than two networks, rather than a pair of networks as is done in most of the conventional global network alignment studies within GONA context. However the two definitions are not inclusive; a multiple network aligner can provide one-to-one mappings as is the case for the Fuse algorithm [29], and a pairwise network aligner can provide non-injective mappings as is the case for the SubMAP algorithm [8]. Since the distinct characteristic of the approaches within GMNA context is outputting many-to-many mappings, that is matched clusters from the networks, where each cluster may contain several proteins from a single species, we choose the many-to-many nomenclature to describe all such approaches. However we formally define the problem in its generality, that is in a form applying to the alignment of two or more networks.

1.5.1 Problem Definition

Given $\prec G_1, G_2, \ldots G_k, S \succ$, where S is a k-partite graph, a *cluster* is defined as the set of nodes of a complete c-partite subgraph of S, for $1 < c \leq k$. For a pair of clusters L_i, L_j, let E_{ij} denote the set of all edges in $E_1, E_2, \ldots, E_k$ with one end point in L_i and the other in L_j. We define the *Global Many-to-Many Network Alignment (GMNA)* problem as a generalization of BGONA as follows:

Definition 1.3 Find maximal set L of non-overlapping clusters of S maximizing:

$$\alpha \times T'(L) + (1 - \alpha) \times H'(L). \tag{1.2}$$

The topological similarity component $T'(L)$ is a generalization of $T(M)$ in Eq. 1.1. For a pair of clusters L_i, L_j, let a denote the number of graphs G_r such that $E_{ij} \cap E_r \neq \emptyset$ and b denote the number of graphs that contain at least one vertex in L_i and at least one vertex in L_j. Intuitively, the latter indicates the maximum number of conserved edges that can be attained between L_i and L_j, whereas the former indicates the actual number of conserved edges between them. Thus *edge conservation rate* between $L_i, L_j \in L$, denoted by t_{ij}, is a/b if $a > 1$ and is 0 otherwise; see Fig. 1.1 for a figure depicting some sample instances. Then $T'(L)$ is defined as the normalized sum of edge conservations between all pairs of clusters, that is, $\sum_{\forall i,j} |E_{ij}| t_{ij} / \sum_{\forall i,j} |E_{ij}|$. Similarly, $H'(L)$ is a generalization of $H(M)$ of Eq. 1.1. Average homological similarity score of a cluster L_i can be defined as the average of a suitable function $w(uv)$, defined on $u, v \in L_i$ such that $uv \in S$, averaged over all such pairs $u, v \in L_i$. Then $H'(L)$ is an average of all the scores over all clusters in L.

Note that this is the definition proposed in [5]. Although the rest of the relevant studies on GMNA including [8, 17, 27, 43, 56, 63] do not provide such a formalized optimization problem definition, it can be deduced from the nature of their proposed algorithms that their optimization goals are similar to that of Eq. 1.2.

1.5.2 Computational Complexity

The following result from [5] states the main computational complexity result regarding the GMNA problem defined in the previous subsection:

Theorem 1.5 *For all $\alpha \neq 0$, the GMNA problem is NP-hard even for the restricted case where all edge weights in S are equal.*

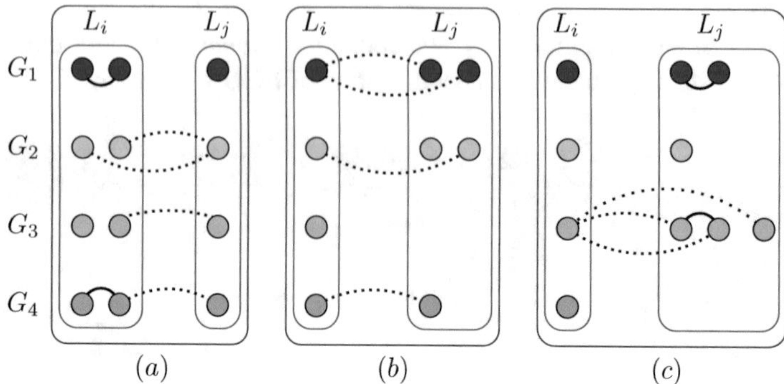

Fig. 1.1 Edge conservation rates on sample GMNA scenarios. The dotted edges represent the protein–protein interactions between the clusters, whereas solid edges represent interactions within a cluster. Nodes of the same PPI network are drawn at the same horizontal layer. (**a**) There are three graphs with edges with one end in L_i and the other end in L_j, thus we have $a = 3$. Note that multiple edges of the same graph count as 1, as is the case for two G_2 edges. There are four graphs with at least one vertex in both L_i and L_j, thus we have $b = 4$. Therefore $t = 3/4$. (**b**) There are three graphs with edges with one end in L_i and the other end in L_j. There are also three graphs with at least one vertex in both L_i and L_j. Thus $t = 3/3$. (**c**) Since there is only one graph with edges with one end in L_i and the other end in L_j, we have $t = 0$

1.5.3 Algorithms

We can categorize the main approaches proposed for GMNA into two: those that employ a GONA solution of pairwise networks as a primitive in their algorithms and those that do not. The former category of GMNA algorithms include [17, 43, 56]. Given k graphs corresponding to PPI networks from k species, IsoRankN [43] starts with the *functional similarity* scores of pairs of nodes from different graphs. This scoring is computed by a random walk on the product graph of every pair of graphs, as in the IsoRank algorithm. These scores give rise to an edge-weighted complete k-partite graph F, where the edge weights correspond to the functional similarity scores. Let S_u denote a heavy star graph centered at u, that is, the set of neighbors v of u in F with weight of uv greater than a threshold. The following procedure is repeated iteratively. One of the input graphs is picked at random, and a heavy star graph is constructed centered at each node in the picked graph. Nodes are ordered in decreasing weights of their star graphs. Starting with the maximum weight one, the star graph is further filtered by excluding the nodes already in output clusters and finding the *minimum conductance* subset within the star graph, that is, subset of vertices that minimizes the ratio of the size of the edge cut-set to the size of the smaller subset after the cut. The resulting minimum conductance subset is merged with subsets centered at other nodes of the same graph, if the star graphs, excluding the centers, show significant overlap. The resulting subgraph is added to the set of output clusters. Similarly SMETANA [56] needs to work with *global correspondence scores* resembling the functional similarity scores of IsoRankN.

The only difference is that the scores employed by SMETANA are produced via a semi-Markov random walk of [55], rather than the random walk of IsoRank. Such a score is further fine-tuned by two different considerations. One updates the score of a pair of nodes based on the scores of their neighbors in their respective networks. The other incorporates the scoring information of overlapping network pairs, so that score of a pair u, v for $u \in G_p, v \in G_q$, is updated by the scores of u, w and w, v, for $w \in G_r, r \neq p \neq q$. Once such scores representing the alignment probability of a pair of nodes is computed, the actual output set of clusters is computed via greedy heuristic by each time considering the pair with the largest alignment probability. If one of the nodes of this pair already belongs to a cluster, the decision to add the other to the same cluster depends on two conditions; the considered pair's score is at least a constant multiple of the average alignment in the cluster and that the cluster size does not exceed a preset bound. SMAL [17] is yet another GMNA algorithm that makes use of other alignment algorithms. Given k graphs, the algorithm chooses a *scaffold graph*, G_s. It then simply takes any alignment algorithm $\mathcal{A}$ suggested for BGONA or GMNA and applies $\mathcal{A}$ on graph pairs G_s, G_i, $\forall i \neq s$. Each output cluster corresponding to $u \in G_s$ is formed by the union of pairwise alignments including u. Although there is no optimization goal as defined in Eq. 1.2, the relevant scores suggested for evaluation purposes are defined similar to that of Eq. 1.2; the conserved interaction score used for evaluations takes into account the interactions between clusters, and the homological similarity score used in evaluations takes into account only proteins within each cluster.

The stand-alone GMNA approaches that do not require the use of another alignment algorithm include [5, 8, 27]. In Beams [5] the edges of the similarity graph are simply the sequence similarities of pairs. Those lower than a threshold are removed. A *backbone* extraction and merge strategy is followed, where a backbone corresponds to the backbone of a cluster consisting of at most one node from each PPI network. Assuming already existing candidates, the best candidate that maximizes the score defined in Eq. 1.2 is chosen as a backbone. Once a backbone is produced, its candidate is constructed by considering the *maximum edge-weighted clique (MEWC)* of its neighborhood graph, which consists of the PPI neighbors of the backbone nodes and their connections in the similarity graph. In the merge step the backbones are greedily merged together. At each iteration, among the *mergeable* groups of pairs of backbones, that is those pairs whose union induces a complete c-partite graph in the similarity graph, the pair maximizing the score in Eq. 1.2 is merged together. In BalanceAli [27], the weight of each edge in the k-partite similarity graph is assigned as a convex combination of local topological similarity and homological similarity. To account for the former, all pairs with sequence similarity below a threshold are eliminated from the similarity graph. Next each remaining edge uv is assigned a weight as a function of the number of conserved edges involving uv, that is number of xy retained in the similarity graph such that uv and xy are PPI graph edges, and the number of conserved triangles, defined similarly. Next the output clusters are formed using a greedy heuristic based on simulated annealing. SubMAP [8] is an algorithm proposed for aligning pathways (directed graphs), but the ideas can be employed directly to align undirected PPI networks as well. Considering its restriction to undirected graphs, it can be treated as a special case of GMNA; it provides one-to-many alignments rather than many-to-many alignments. It provides pairwise alignments, as compared to aligning more than two networks achieved by the previously described GMNA algorithms. In terms of the formal optimization goal, it is similar to that of Eq. 1.2, in the sense that the homological similarity is computed within each cluster (one-to-many) and the conserved interactions are counted as those induced by a pair of clusters. The algorithm constructs all connected subnetworks of size at most k in both networks. A *combined similarity* score for each pair of subnetworks, one from G_1 and one from G_2, so that one subnetwork consists only of a single node, is computed via a random walk-type procedure aimed at a function similar to that of Eq. 1.2. Considering each pair of networks as a node and creating an edge between two nodes if the subnetworks are not disjoint, a graph similar in function to the conflict graph described in Sect. 1.4 is constructed. Finally a maximum weight-independent set heuristic is applied on this conflict graph to extract the output one-to-many clusters.

1.6 Network Alignment Evaluation Criteria

We provide a discussion of two types of criteria: those with no biological connotation and those with biological connotation. For the former, the employed evaluation procedure or metric makes use only of the input instance, such as input PPI networks and homological similarity scores, and the output alignment. The latter employs further biological data, such as functional pathway, and known gene annotations, for evaluating the alignment.

1.6.1 Procedures and Metrics with No Biological Connotation

One category of network alignment evaluation metrics includes those defined within the context of the formal optimization goals determined with respect to the specific alignment problem definition. The topological similarity denoted by the function T in the Definition 1.1, 1.2, and 1.3 is one such metric. Given an input instance $\prec G_1, G_2, S \succ$ and an output alignment M, the particular topological similarity definitions that can be used for the evaluations of BGONA and CGONA algorithms include the edge conservation score EC, that is, $|\{(ux, vy) \in E_1 \times E_2 | uv, xy \in M \vee uy, xv \in M\}|$. In some cases such a score is used after some appropriate normalizations. In the ICS metric, assuming $|V_1| \leq |V_2|$, EC is normalized by the size of the subgraph of G_2 induced by the nodes incident on edges in M [49]. For the S^3 metric, the difference between $|E_1|$ and the size of the conserved edge set is further added to the normalization factor [58]. Another useful metric is the size of the largest connected component of the alignment graph (graph composed of edges that contribute to EC). Likewise, for the topological similarity of the GMNA studies, the metrics employed in the definition of the T' function can be employed.

Different from the BGONA and CGONA settings, the GMNA evaluations include a *coverage* metric as well. It indicates the total number of proteins covered by the output clusters. Although not meaningful for the one-to-one alignments, it is a useful metric for the many-to-many alignments, as these numbers may vary with respect to the employed alignment algorithm. High coverage alignments leave out much less unexplained data by proposing orthology relations in the form of clusters for most of the proteins. The average size of the output clusters may also be important, especially if the employed evaluation metrics with biological connotation disregard such a property. Thus coverage is usually accompanied with the number of output clusters among useful statistics.

Self-alignment [45, 49] and *error tolerance* [61] refer to similar evaluation procedures that can be used to measure the robustness of an alignment algorithm under the BGONA and CGONA settings. An instance $\prec, G_1, G_2, S \succ$ is created so that G_1 is an existing high-confidence PPI network of some species and G_2 is a noisy version of G_1. The artificial noise can be introduced by either random re-wirings of the interactions [61] in G_1 or by inserting a certain percentage of low-confidence interactions into high-confidence G_1 [45, 49]. In such a case, since the correct node mappings are already known, additional metrics called *node correctness (NC)* and *interaction correctness (IC)* are also employed. The former refers to the number of nodes of G_2 that are mapped to their correct counterparts in G_1, whereas the latter refers to the number of conserved interactions contributing to EC that are between a pair of correct node mappings. Synthetically generated networks have found use in two other ways in global network alignment evaluations. First, synthetic networks produced according to some proper growth model, such as those described in NAPABench [54], are used in both BGONA and GMNA studies, where networks of the same family are aligned and the produced alignments are compared with respect to measures including NC, IC [5, 16, 54]. Second, they have found use in finding the statistical significance of alignments of actual PPI networks in BGONA setting [41]. In this scenario, assuming a proper null model of random network generation for PPI networks, the alignment of actual PPI networks is compared against that of synthetic networks generated with the null model to measure the significance of the former alignment.

Depending on the exact problem definition and the underlying optimization function, a major issue in the global network alignment has been the computational intractability of the problem versions; almost all the appropriate optimization formulations are computationally hard. It becomes even more apparent with some input PPI networks containing tens of thousands of nodes and interactions. An important feature expected of the global network alignments is then scalability; the running-time performances of the suggested methods should not degrade drastically with increasing network sizes. Thus yet another important evaluation criteria are the worst-case running-time bounds of the algorithms and the required actual CPU times on real PPI networks.

1.6.2 Procedures and Metrics with Biological Connotation

These procedures and metrics are designed so as to validate the likelihood that the output alignment pairs (in case of BGONA and CGONA) or clusters (in case of GMNA) provide actual orthology relationships and functional similarities.

One common such validation involves gene ontology annotations, such as those provided in the GO Consortium [7]. Regarding one-to-one settings, a definition of *GO Consistency (GOC)* can be employed, where the contribution of an alignment pair $uv \in M$ to GOC is $|GO(u) \cap GO(v)|/|GO(u) \cup GO(v)|$. Here, $GO(x)$ denotes the set of GO terms annotating a protein x. Two potential problems may arise if such an implementation is implemented directly. First, since many GO annotations themselves are based on sequence alignments, such comparisons may produce misleading results when an algorithm emphasizes homological similarities in its alignments. The second potential problem is due to the fact that GO categories are organized hierarchically as a large directed acyclic graph (DAG). Overlaps in categories close to the root of the hierarchy are insignificant as the categories are too general. On the contrary, there might be very specific categories at levels further away from the root of the GO DAG, and expecting exact category overlaps can be a strong requirement for GO consistency evaluations. To handle the problems of the first type, one alternative is to compare the output alignments of different algorithms, under fixed total sequence similarity scores, when possible [3, 66]. Another alternative is to limit the GO categories used for evaluation to those that are not based on BLAST evidence alone. Limiting the GO categories to the evidence codes IPI, IGI, IMP, IDA, IEP, TAS, and IC is one such possibility [3, 42]. To resolve the problems of the second type, usually not all the GO DAG is taken into account and the GO categories under consideration are standardized. This is done by restricting the protein annotations to level five of the GO DAG by ignoring the higher-level annotations and replacing the deeper-level category annotations with their ancestors at the restricted level [3, 43, 62].

Gene ontology-based evaluations extend to the many-to-many setting as well. Let an *annotated* cluster indicate one that contains at least two proteins annotated with some GO categories, assuming that the previously described standard GO categories are employed for similar considerations. For an annotated cluster L_i, let $GO_{\cap}(L_i)$ and $GO_{\cup}(L_i)$ indicate, respectively, the intersection and the union sets of GO annotations of proteins in L_i. The extension of GOC in the one-to-one settings is the normalized GOC score, $nGOC$, which is defined as the weighted mean of $|GO_{\cap}/GO_{\cup}|$ over all annotated clusters, where the weight of each cluster is the number of annotated proteins it contains. *Mean normalized entropy (MNE)* is also a GO-based evaluation metric. The normalized entropy of an annotated cluster L_i is defined as $NE(L_i) = -\frac{1}{\log d} \times \sum_{r=1}^{d} p_r \times \log p_r$, where p_r is the fraction of proteins in L_i with the annotation GO_r, and d represents the number of different GO annotations in L_i. For MNE, the sum of these values is averaged over the total number of annotated clusters. Note that lower MNE values indicate better alignments in terms of this measure. Relevant specificity and sensitivity metrics on alignments can also be defined in the GMNA setting. An annotated cluster is considered *consistent* if all of its proteins share at least one common standard GO annotation. *Specificity* of a GMNA is defined as the ratio of consistent clusters to annotated clusters. For a given GO category, let its *closest cluster* denote the cluster that contains the maximum number of proteins annotated with this GO category. The *sensitivity* of an alignment is then defined as the average, over all GO categories, of the fraction of aligned nodes annotated with a GO category that are also in its closest cluster.

1.7 Discussion

This chapter provides an introduction to the global PPI network alignment and a review of different problem versions and the main approaches proposed for them. Although much excellent work has been published in the relevant literature, there still remain directions that can be explored further. One of the relatively less studied versions of the problem is CGONA. There are many open problems within this context, which could especially be of interest to theoretical computer scientists, as the combinatorial properties of the conflict graphs under different settings of k_1, k_2 are still unknown.

Systematic application of network alignments in functional orthology predictions is another open research problem. Although most of the existing network alignment studies suggest protein function predictions via *annotation transfers*, that is, via assigning the annotations of a protein in an aligned pair (or cluster in GMNA) to the unannotated member of the same pair (or cluster), a detailed analysis demonstrates that such automated transfers by themselves may not always be sufficient to provide immediate function predictions. Incorporating the global alignment results into the function prediction methods using network analysis techniques provides more reliable predictions [59]. A methodological treatment of the issue along this direction is still open.

Finally, PPI network prediction is an application area where network alignment may be of use, although it has not received as much attention as functional orthology. Interaction prediction methods usually limit the scope of their studies to a single network, employing data based on genomic context, structure, domain, sequence information, or existing network topology. Incorporating multiple species network data for PPI network prediction entails the design of novel models encompassing both network reconstruction and network alignment, since the goal of network alignment is to provide functionally orthologous proteins from multiple networks and such orthology information can be used in guiding interolog transfers. However, such an approach raises the classical chicken or egg problem; alignment methods assume error-free networks, whereas network prediction via interolog transfers works effectively if the functionally orthologous proteins are determined with high precision. Such intertwinements exist in other computer science areas. The *simultaneous localization and mapping (SLAM)* problem from Robotics is one example that has received considerable attention [12]. The goal is to estimate the pose of a robot and the map of the environment at the same time. A similar problem arises, since a map is needed for localization and a good pose estimate is needed for mapping. Employing network alignments in PPI network predictions has been investigated previously, in a few studies [6, 44]. L-GRAAL, although does not provide a comprehensive methodological view, proposes to use network alignments in interaction predictions. A problem analogous to the SLAM, the problem of *simultaneous prediction and alignment of networks (SPAN)*, has been proposed in [6], and the problem is studied in its simultaneous nature. Further work along this direction could prove useful for researchers in the areas of network alignment and network prediction.

References

1. Abaka, G., Biyikoglu, T., Erten, C.: Campways: constrained alignment framework for the comparative analysis of a pair of metabolic pathways. Bioinformatics **29**(13), i145–i153 (2013)
2. Aebersold, R., Mann, M.: Mass spectrometry-based proteomics. Nature **422**(6928), 198–207 (2003)
3. Aladağ, A.E., Erten, C.: Spinal: Scalable protein interaction network alignment. Bioinformatics **29**(7), 917–924 (2013)
4. Alkan, F., Bíyíkoglu, T., Demange, M., Erten, C.: Structure of conflict graphs in constraint alignment problems and algorithms. Discrete Math. Theore. Comput. Sci. **21**(4), (2019). http://dmtcs.episciences.org/5755
5. Alkan, F., Erten, C.: Beams: backbone extraction and merge strategy for the global many-to-many alignment of multiple ppi networks. Bioinformatics **30**(4), 531–539 (2014)
6. Alkan, F., Erten, C.: Sipan: Simultaneous prediction and alignment of protein-protein interaction networks. Bioinformatics (Oxford, England) **31**, 2356–2363 (2015)
7. Ashburner, M., Ball, C.A., Blake, J.A., et al.: Gene ontology: tool for the unification of biology. Nat. Genet. **25**(1), 25–29 (2000)
8. Ay, F., Kellis, M., Kahveci, T.: Submap: aligning metabolic pathways with subnetwork mappings. J. Comput. Biol. **18**(13), 219–235 (2011)
9. Bader, G.D., Hogue, C.W.V.: Analyzing yeast protein-protein interaction data obtained from different sources. Nat. Biotechnol. **20**(10), 991–997 (2002)
10. Bandyopadhyay, S., Sharan, R., Ideker, T.: Systematic identification of functional orthologs based on protein network comparison. Genome Research **16**(3), 428–35 (2006)
11. Bayati, M., Borgs, C., Chayes, J.T., Zecchina, R.: Belief propagation for weighted b-matchings on arbitrary graphs and its relation to linear programs with integer solutions. SIAM J. Discrete Math. **25**(2), 989–1011 (2011)
12. Cadena, C., Carlone, L., Carrillo, H., Latif, Y., Scaramuzza, D., Neira, J., Reid, I., Leonard, J.J.: Past, present, and future of simultaneous localization and mapping: Toward the robust-perception age. Trans. Rob. **32**(6), 1309–1332 (2016). https://doi.org/10.1109/TRO.2016.2624754
13. Chindelevitch, L., Liao, C.S., Berger, B.: Local optimization for global alignment of protein interaction networks. In: Pacific Symposium on Biocomputing, pp. 123–132 (2010)
14. Chindelevitch, L., Ma, C.Y., Liao, C.S., Berger, B.: Optimizing a global alignment of protein interaction networks. Bioinformatics **29**(21), 2765–2773 (2013)
15. Clark, C., Kalita, J.: A comparison of algorithms for the pairwise alignment of biological networks. Bioinformatics (Oxford, England) **30**, 2351–2359 (2014)

16. Clark, C., Kalita, J.: A multiobjective memetic algorithm for PPI network alignment. Bioinformatics **31**(12), 1988–1998 (2015)
17. Dohrmann, J., Puchin, J., Singh, R.: Global multiple protein-protein interaction network alignment by combining pairwise network alignments. BMC Bioinformatics **16**(13), S11 (2015)
18. Dutkowski, J., Tiuryn, J.: Identification of functional modules from conserved ancestral protein–protein interactions. Bioinformatics **23**(13), i149–i158 (2007)
19. Edmonds, J.: Maximum matching and a polyhedron with 0, 1-vertices. J. Res. Natl. Bur. Stand. B **69**, 125–130 (1965)
20. El-Kebir, M., Heringa, J., Klau, G.W.: Lagrangian relaxation applied to sparse global network alignment. In: Loog, M., Wessels, L., Reinders, M.J.T., de Ridder, D. (eds.) Pattern Recognition in Bioinformatics. pp. 225–236. Springer, Berlin, Heidelberg (2011)
21. Elmsallati, A., Clark, C., Kalita, J.: Global alignment of protein-protein interaction networks: A survey. IEEE/ACM Trans. Comput. Biol. Bioinform. **13**(4), 689–705 (2016). https://doi.org/10.1109/TCBB.2015.2474391
22. Fertin, G., Rizzi, R., Vialette, S.: Finding occurrences of protein complexes in protein–protein interaction graphs. J. Discrete Algorithms **7**(1), 90–101 (2009)
23. Finley, R.L., Brent, R.: Interaction mating reveals binary and ternary connections between drosophila cell cycle regulators. Proc. Natl. Acad. Sci. USA **91**(26), 12980–12984 (1994)
24. Fionda, V.: Protein-protein interaction network alignment: Algorithms and tools. In: Pan, Y., Wang, J., Li, M. (eds.) Algorithmic and Artificial Intelligence Methods for Protein Bioinformatics, chap. 22, pp. 431–448. Wiley, Hoboken, NJ (2013)
25. Flannick, J., Novak, A., Srinivasan, B.S., McAdams, H.H., Batzoglou, S.: Graemlin: general and robust alignment of multiple large interaction networks. Genome Research **16**(9), 1169–1181 (2006)
26. Gabow, H.N.: Scaling algorithms for network problems. In: Proceedings of the 24th Annual Symposium on Foundations of Computer Science. pp. 248–258. SFCS '83, IEEE Computer Society, Washington, DC, USA (1983). https://doi.org/10.1109/SFCS.1983.68
27. Gao, J., Song, B., Ke, W., Hu, X.: Balanceali: Multiple PPI network alignment with balanced high coverage and consistency. IEEE Trans. NanoBiosci. **16**(5), 333–340 (2017). https://doi.org/10.1109/TNB.2017.2705521
28. Garey, M.R., Johnson, D.S.: Computers and Intractability: A Guide to the Theory of NP-Completeness. W. H. Freeman (1979)
29. Gligorijević, V., Malod-Dognin, N., Pržulj, N.: Fuse: multiple network alignment via data fusion. Bioinformatics **32**(8), 1195–1203 (2015)
30. Goh, C.S., Cohen, F.E.: Co-evolutionary analysis reveals insights into protein-protein interactions. J. Mol. Biol. **324**(1), 177–192 (2002)
31. Golub, G.H., Van Loan, C.F.: Matrix Computations (3rd edn.). Johns Hopkins University Press, Baltimore, MD, USA (1996)
32. Guzzi, P., Milenković, T.: Survey of local and global biological network alignment: The need to reconcile the two sides of the same coin. Brief. Bioinform. **19** (01 2017)
33. Han, J.D.J., Bertin, N., Hao, T., Goldberg, D.S., Berriz, G.F., Zhang, L.V., Dupuy, D., Walhout, A.J., Cusick, M.E., Roth, F.P., Vidal, M.: Evidence for dynamically organized modularity in the yeast protein–protein interaction network. Nature **430**(6995), 88–93 (2004)
34. Hashemifar, S., Ma, J., Naveed, H., Canzar, S., Xu, J.: ModuleAlign: module-based global alignment of protein–protein interaction networks. Bioinformatics **32**(17), i658–i664 (08 2016)
35. Hashemifar, S., Xu, J.: Hubalign: An accurate and efficient method for global alignment of protein-protein interaction networks. Bioinformatics (Oxford, England) **30**, i438–i444 (2014). https://doi.org/10.1093/bioinformatics/btu450
36. Hunter, H.B., Aaron, A.E., Lars, L.M., Curt, C., Marcus, M.W.: Evolutionary rate in the protein interaction network. Science **296**(5568), 750–752 (2002)
37. Kazemi, E., Hassani, H., Grossglauser, M., Pezeshgi Modarres, H.: Proper: global protein interaction network alignment through percolation matching. BMC Bioinformatics **17**(1), 527 (Dec 2016)

38. Kelley, B.P., Sharan, R., Karp, R.M., Sittler, T., Root, D.E., Stockwell, B.R., Ideker, T.: Conserved pathways within bacteria and yeast as revealed by global protein network alignment. Proc. Natl. Acad. Sci. **100**(20), 11394–11399 (2003)
39. Kelley, B.P., Yuan, B., Lewitter, F., Sharan, R., Stockwell, B.R., Ideker, T.: Pathblast: a tool for alignment of protein interaction networks. Nucleic Acids Res. **32**(Web-Server-Issue), 83–88 (2004)
40. Koyutürk, M., Kim, Y., Topkara, U., Subramaniam, S., Szpankowski, W., Grama, A.: Pairwise alignment of protein interaction networks. J. Comput. Biol. **13**(2), 182–199 (2006)
41. Kuchaiev, O., Milenković, T., Memišević, V., Hayes, W., Pržulj, N.: Topological network alignment uncovers biological function and phylogeny. J. R. Soc. Interface **7**(50), 1341–1354 (2010)
42. Kuchaiev, O., Pržulj, N.: Integrative network alignment reveals large regions of global network similarity in yeast and human. Bioinformatics **27**(10), 1390–1396 (2011)
43. Liao, C.S., Lu, K., Baym, M., Singh, R., Berger, B.: Isorankn: spectral methods for global alignment of multiple protein networks. Bioinformatics **25**(12), i253–i258 (2009)
44. Malod-Dognin, N., Pržulj, N.: L-GRAAL: Lagrangian graphlet-based network aligner. Bioinformatics **31**(13), 2182–2189 (2015)
45. Milenković, T., Leong Ng, W., Hayes, W., Pržulj, N.: Optimal network alignment with graphlet degree vectors. Cancer Inform. **9**, 121–137 (2010)
46. Mohammadi, S., Grama, A.: Biological network alignment. In: Koyutürk, M., Subramaniam, S., Grama, A. (eds.) Functional Coherence of Molecular Networks in Bioinformatics, chap. 5, pp. 97–136. Springer (2011)
47. Narayanan, M., Karp, R.M.: Comparing protein interaction networks via a graph match-and-split algorithm. J. Comput. Biol. **14**(7), 892–907 (2007)
48. Neyshabur, B., Khadem, A., Hashemifar, S., Arab, S.S.: NETAL: a new graph-based method for global alignment of protein–protein interaction networks. Bioinformatics **29**(13), 1654–1662 (2013)
49. Patro, R., Kingsford, C.: Global network alignment using multiscale spectral signatures. Bioinformatics **28**(23), 3105–3114 (2012)
50. Phan, H.T.T., Sternberg, M.J.E.: PINALOG: a novel approach to align protein interaction networks–implications for complex detection and function prediction. Bioinformatics **28**(9), 1239–1245 (2012)
51. Raymond, J.W., Willett, P.: Maximum common subgraph isomorphism algorithms for the matching of chemical structures. J. Comput. Aided Mol. Des. **16**(7), 521–533 (2002)
52. Remm, M., Storm, C.E., Sonnhammer, E.L.: Automatic clustering of orthologs and in-paralogs from pairwise species comparisons. J. Mol. Biol. **314**(5), 1041–1052 (2001)
53. Remm, M., Storm, C., Sonnhammer, E.: Automatic clustering of orthologs and in-paralogs from pairwise species comparisons. J. Mol. Biol. **314**, 1041–52 (2002). https://doi.org/10.1006/jmbi.2000.5197
54. Sahraeian, S., Yoon, B.J.: A network synthesis model for generating protein interaction network families. PLoS ONE **7**, e41474 (2012)
55. Sahraeian, S., Yoon, B.J.: Resque: Network reduction using semi-Markov random walk scores for efficient querying of biological networks. Bioinformatics (Oxford, England) **28**, 2129–2136 (2012)
56. Sahraeian, S.M.E., Yoon, B.J.: Smetana: Accurate and scalable algorithm for probabilistic alignment of large-scale biological networks. PLoS ONE **8**(7), e67995 (2013)
57. Sakai, S., Togasaki, M., Yamazaki, K.: A note on greedy algorithms for the maximum weighted independent set problem. Discrete Appl. Math. **126**(2-3), 313–322 (2003). https://doi.org/10.1016/S0166-218X(02)00205-6
58. Saraph, V., Milenković, T.: MAGNA: Maximizing accuracy in global network alignment. Bioinformatics **30**(20), 2931–2940 (07 2014)
59. Sharan, R., Ideker, T.: Modeling cellular machinery through biological network comparison. Nature Biotechnology **24**(4), 427–433 (2006)

60. Sharan, R., Suthram, S., Kelley, R.M., Kuhn, T., McCuine, S., Uetz, P., Sittler, T., Karp, R.M., Ideker, T.: Conserved patterns of protein interaction in multiple species. Proc. Natl. Acad. Sci. USA **102**(6), 1974–1979 (2005)
61. Singh, R., Xu, J., Berger, B.: Pairwise global alignment of protein interaction networks by matching neighborhood topology. In: Speed, T., Huang, H. (eds.) Research in Computational Molecular Biology. pp. 16–31. Springer, Berlin, Heidelberg (2007)
62. Singh, R., Xu, J., Berger, B.: Global alignment of multiple protein interaction networks. In: Pacific Symposium on Biocomputing, pp. 303–314 (2008)
63. Singh, R., Xu, J., Berger, B.: Global alignment of multiple protein interaction networks with application to functional orthology detection. Proc. Natl. Acad. Sci. **105**(35), 12763–12768 (2008). https://doi.org/10.1073/pnas.0806627105
64. Tian, W., Samatova, N.: Pairwise alignment of interaction networks by fast identification of maximal conserved patterns. Pac. Symp. Biocomput. **14**, 99–110 (2009)
65. Viterbi, A.: Error bounds for convolutional codes and an asymptotically optimum decoding algorithm. IEEE Trans. Inf. Theor. **13**(2), 260–269 (2006). https://doi.org/10.1109/TIT.1967.1054010
66. Zaslavskiy, M., Bach, F.R., Vert, J.P.: Global alignment of protein-protein interaction networks by graph matching methods. Bioinformatics **25**(12), 259–267 (2009)

Chapter 2
Effective Random Walk Models for Comparative Network Analysis

Hyundoo Jeong

Abstract Comparative network analysis provides an effective computational means to identify the similarities and differences between biological networks so that it helps transferring the prior knowledge across different biological networks. Since identifying the optimal biological network alignment is practically infeasible due to the computational complexity, a number of heuristic network alignment algorithms have been proposed. Among various heuristic approaches, comparative network analysis methods using random walk models are very effective to construct a reliable network alignment because it can accurately estimate the node correspondence by integrating node similarity and topological similarity. In this chapter, we introduce effective random walk models and their applications in developing comparative network analysis algorithms.

Keywords Comparative network analysis · Network alignment · Random walk · Node correspondence · Steady-state probability

2.1 Introduction

Comparative network analysis algorithms aim to identify the node (or subnetwork) mapping between different biological networks so that it can provide the effective computational means to exchange the prior knowledge among different biological networks [1]. For example, the function of a protein in the less-studied protein–protein interactions (PPI) network can be predicted based on the function of the matching protein in the well-studied PPI network. Additionally, the comparative network analysis enables biological experiments targeting particular genes or proteins in a model species, where it cannot be directly applied to human because of ethical issues. As a result, it expedites deciphering and predicting the complex

H. Jeong (✉)
Department of Mechatronics Engineering, Incheon National University, Incheon, South Korea
e-mail: hdj@inu.ac.kr

© Springer Nature Switzerland AG 2021
B.-J. Yoon, X. Qian (eds.), *Recent Advances in Biological Network Analysis*,
https://doi.org/10.1007/978-3-030-57173-3_2

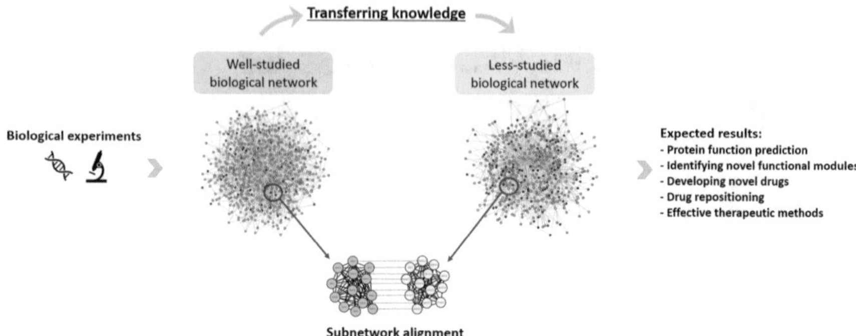

Fig. 2.1 Illustration of comparative network analysis and potential applications in biomedical research

biological mechanisms in a human based on the experimental results of model species. Ultimately, comparative network analysis algorithms can be utilized to accelerate the development of novel drug and effective therapy for complex diseases such as cancer and Alzheimer's disease because it helps to save cost and time for biological experiments to validate biological functions of proteins (Fig. 2.1).

Comparative network analysis can be broadly classified into three categories: (1) global network alignment, (2) local network alignment, and (3) network querying. The global network alignment aims to identify the best global node mapping between different networks, where it can maximize the overall similarity of aligned nodes. The goal of the local network alignment is identifying the conserved subnetwork regions in different networks. The network querying identifies the subnetworks in a target network that are similar to a given query network in terms of topology as well as biological functions. Typically, the number of nodes in a target network is much larger than the query network. The comparative network analysis algorithms can be further categorized based on the number of networks to be compared. Although pairwise network alignment algorithms can compare only two biological networks, multiple network alignment algorithms can identify the conserved subnetworks across more than two networks so that it can help understanding more deeper biological insights conserved in multiple species. However, multiple network alignment algorithms should solve the scalability and computational complexity issues in a practical point of view.

Various network alignment algorithms have been proposed based on different heuristic strategies because the network alignment problem can be mathematically considered as the subgraph isomorphism problem, which is NP-complete. Additionally, when considering the size of real-world biological networks, deriving the mathematically optimal network alignment would be infeasible due to the computational complexity. IsoRank [2] utilizes the modification of PageRank algorithm to estimate the node correspondence for network pairs and then it constructs the one-to-one node mapping based on a modified greedy algorithm. IsoRankN [3] extends IsoRank to deal with multiple networks and utilizes the spectral clustering called PageRank-Nibble [4] to yield the final network alignment results. GRAAL [5] analyzes the graphlet, where it is the generalized node degree, and constructs the pairwise network alignment based on the similarity of graphlet degree signature for each node. Several variations of GRAAL have been proposed [5–8]. GHOST [9] proposes a novel metric called induced conserved structure that can measure the topological similarity and utilizes the seed-and-extension approach to yield the best network alignment results. BEAMS [10] first extracts the minimal set of disjoint cliques called backbone and then iteratively merges backbones to yield the final alignment that can maximize the overall alignment scores. FUSE [11] utilizes nonnegative matrix trifactorization to yield the accurate alignment results. NetCoffee [12] identifies all possible combinations of bipartite graphs and then refines the edges using the protein sequence similarity and topological similarity. Then, among candidate edges in the bipartite graphs, it merges the selected edges in the bipartite graph through the simulated annealing. HubAlign [13] first computes the weights of nodes and edges based on their topological significance (i.e., probability to be a hub), and then it computes the alignment scores using the weight of node and edges. Finally, it constructs a global alignment via greedy seed-and-extension approach. SANA [14] defines the objective function using S^3 score and utilizes the simulated annealing to obtain the optimal pairwise network alignment. MAGNA [15] and MAGNA++ [16] define the cost function based on the linear combination of node and edge conservation measurement and employ the genetic algorithm to derive the optimal alignment results.

Recently, novel random walk models for biological network alignment have been proposed [17–19]. In fact, a random walk model for analyzing a network (or graph) has been gaining its popularity in various fields as PageRank algorithm based on a random walk approach has been successfully applied to Google's searching engine [20]. Especially, in a biological network alignment problem, the random walk-based algorithms are very effective because of their distinctive advantages. First, the random walk models have an inherent flexibility to model the topological variations such as node insertions and deletions. Moreover, although the accurate estimation of node correspondence is a pivotal procedure to yield reliable network alignments, the structural changes among different biological networks make it difficult to integrate different types of similarities in a balanced manner. However, estimating the node correspondence through a random walk model is not limited to the topological changes. Additionally, since the network can be represented as the adjacency matrix, the random walk-based network alignment algorithms take an advantage of various mathematical theories and it requires a low computational cost as the adjacency matrix representing a biological network is typically sparse. In this chapter, we introduce the effective random walk models to estimate node correspondence and their application in global network alignment algorithms for protein–protein interactions networks.

2.2 Methods

Suppose that we have a pair of PPI networks and it can be modeled as a graph $G_X = (U, D)$ and $G_Y = (V, E)$, where a node represents a protein in each PPI network (i.e., $u_i \in U$ and $v_j \in V$) and an edge $d_{i,j}$ (or $e_{i,j}$) indicates interactions of two proteins u_i and u_j (or v_i and v_j), and the weight of edge can represent the strength or reliability of protein interactions. When comparing PPI networks, the node similarity is usually available by estimating the similarity of amino-acids sequence for a pair of proteins in different PPI networks. Although the most network alignment algorithms utilize the BLAST bit score (or e-value) as the node similarity [10, 14, 17], other types of similarities or their combinations can be utilized if available [6, 21].

2.2.1 *Problem Formulation*

The goal of global PPI network alignment is identifying one-to-one (or many-to-many) mapping between nodes in different PPI networks. One popular criterion to construct an accurate network alignment is the maximum expected accuracy (MEA) criterion, which aims to maximize the expected number of correctly aligned nodes. In fact, the MEA approach proves its effectiveness in sequence alignment algorithms [22–26]. Suppose that there is a true alignment A^* for pairwise networks, where it is typically unknown and need to be identified. Then, given a network alignment A, the accuracy can be defined as follows:

$$Accuracy\left(A, A^*\right) = \frac{1}{|A|} \sum_{(u_i \sim v_j) \in A} \mathbf{1}\left(u_i \sim v_j \in A^*\right), \tag{2.1}$$

where $\mathbf{1}(\cdot)$ is an indicator function that results 1 if the node mapping $u_i \sim v_j$ is in the true alignment A^*, otherwise $\mathbf{1}(\cdot)$ becomes 0. Since we cannot know the true alignment A^*, the acceptable alternative approximation for the accuracy would be the expected accuracy, which is given by

$$\mathbf{E}_{A^*}\left[Accuracy\left(A, A^*\right)\right] = \frac{1}{|A|} \sum_{(u_i \sim v_j) \in A} P\left[u_i \sim v_j \mid G_X, G_Y\right], \tag{2.2}$$

where $P\left[u_i \sim v_j \mid G_X, G_Y\right]$ is the node alignment probability. Note that we consider the node alignment probability as the node correspondence and interchangeably utilize two terms depending on the context. The goal of global network alignment is identifying the pairwise node mapping that results the maximum expected accuracy (MEA), where it is given by

$$A^* = \arg\max_{A} \sum_{\forall(u_i \sim v_j) \in A} P\left[u_i \sim v_j \mid G_X, G_Y\right]. \tag{2.3}$$

Based on the MEA criteria, the essential first step to obtain quality network alignment is an accurate estimation of node correspondence (i.e., node alignment probability $P\left[u_i \sim v_j \mid G_X, G_Y\right]$). One popular approach to estimate node correspondence is computing the universal similarity by integrating the topological similarity and node-level similarity because it has been shown that the orthologous proteins having similar biological functions could have high compositional similarity and similar interaction patterns to its neighboring proteins [27, 28]. Hence, it is important to integrate the node similarity (e.g., amino-acids sequence similarity, protein function, or composition) and topological similarity (i.e., interaction patterns to its neighboring nodes) in a balanced manner to yield an accurate estimate of node correspondence. Although there are conserved proteins that can be mapped to each other across different PPI networks [27, 28], biological heterogeneity results inserted or deleted proteins in different PPI networks and these topological changes make it challenging to effectively integrate two different types of similarities. Here, we introduce three random walk models, where it can effectively integrate two different similarities even though there are topological variations, so that it can accurately estimate the node correspondence. Then, we will present their applications in pairwise global network alignment algorithms.

2.2.2 *Semi-Markov Random Walk Model*

Given a network $G = (V, E)$, suppose that the random walker can move to its neighboring node at each time and the transition probability of the random walker from the node v_m to the node v_n is given by $p\left(v_n \mid v_m\right) = \frac{w(v_m, v_n)}{\sum_{v_n \in \mathcal{N}(v_m)} w(v_m, v_n)}$, where $w\left(v_m, v_n\right)$ is the weight of edge connecting the node v_m and v_n, and $\mathcal{N}\left(v_m\right)$ is a set of neighboring nodes of v_m. In a typical Markov random walk model, a random walker spends a constant amount of time at each position before changing its position. However, in a semi-Markov random walk model (SMRW), the random walker can stay for the random amount of time at each position before performing a transition.

To estimate the node correspondence through the semi-Markov random walk model, it considers the simultaneous semi-Markov random walk over a pair of networks and it imposes a sophisticated random walk rule, where the expected staying time of the random walkers at each pair of nodes is proportional to the node similarity $s\left(u_i, v_j\right)$ [17]. According to the rule, the random walker simultaneously stays a longer time at the node pair u_i and v_j if they have higher node similarity as well as similar interaction patterns to their neighboring nodes. Through the random walk protocol, the node correspondence can be estimated through the long-run proportion of time that the random walker simultaneously stays at the pair of node, where it can be estimated through a steady-state probability of the random walkers (see Ref. [17] for details). Then, the node correspondence between a pair of nodes in different networks is given by

$$c(u_i, v_j) = \frac{\pi_X(u_i) \cdot \pi_Y(v_j) \cdot s(u_i, v_j)}{\sum_{i'=1}^{N_X} \sum_{j'=1}^{N_Y} \pi_X(u_{i'}) \cdot \pi_Y(v_{j'}) \cdot s(u_{i'}, v_{j'})}, \tag{2.4}$$

where N_X (or N_Y) is the number of nodes in the network G_X (or G_Y), and $\pi_X(u_i)$ and $\pi_Y(v_j)$ the steady-state probability of the random walker on the network G_X and G_Y, respectively.

2.2.3 Context-Sensitive Random Walk Model

The context-sensitive random walk model for network alignment is motivated by the pair-hidden Markov model (HMM) [29], where it has been widely utilized for a comparative analysis of biological sequences that yields a biological sequence alignment, because there are conceptual similarities between a biological network alignment and sequence alignment (see Note 1).

The context-sensitive random walk (CSRW) model considers the simultaneous random walk over two biological networks, and the random walkers can adjust their mode of walk depending on the context at the current position of the random walkers [18]. That is, if there are a pair of neighboring nodes with a positive node similarity, both random walkers in different biological networks can simultaneously move to their neighboring nodes. The simultaneous movement of the random walkers conceptually corresponds to the aligned nodes in different networks. The transition probability for the simultaneous walk is given by

$$P\left[(u_i, v_j) \mid (u_c, v_c)\right] = \frac{s(u_i, v_j)}{\sum_{(u_{i'}, v_{j'}) \in \mathcal{N}(u_c, v_c)} s(u_{i'}, v_{j'})}. \tag{2.5}$$

After examining the context of the current position, if there are no neighboring nodes with a positive node similarity, the CSRW model randomly selects one of the random walkers and it allows the selected walker to perform a transition to its neighboring node and the other random walker stays at the current position so that it can search the potential matching node pair based on the updated context. The individual walk can model the inserted (or deleted) nodes, and the transition probabilities of each random walker are given by

$$P\left[(u_i, v_c) \mid (u_c, v_c)\right] = \frac{|N_X|}{|N_X| + |N_Y|} \cdot \frac{1}{\mathcal{N}(u_c)} \tag{2.6a}$$

and

$$P\left[\left(u_c, v_j\right) \mid (u_c, v_c)\right] = \frac{|N_Y|}{|N_X| + |N_Y|} \cdot \frac{1}{\mathcal{N}(v_c)}, \tag{2.6b}$$

where $|N_X|$ (or $|N_Y|$) is the number of nodes in the network G_X (or G_Y).

Through the CSRW model, the node correspondence between a protein pair (u_i, v_j) can be estimated based on the long-run proportion of time that the random walkers simultaneously enter to the node pair and it can be effectively estimated through the steady-state probability of the random walkers and the following equation:

$$c\left(u_i, v_j\right) = \sum_{(u_p, v_p) \in \mathcal{N}(u_i, v_j)} \pi\left(u_p, v_q\right) P\left[\left(u_i, v_j\right) \mid \left(u_p, v_p\right)\right], \tag{2.7}$$

where $c\left(u_i, v_j\right)$ is the node correspondence for the node pair u_i and v_j, and $\pi\left(u_p, v_q\right)$ is the steady-state probability of the random walkers (i.e., the staying time of the random walkers at the node pair u_p and v_q).

2.2.4 CUFID Model

The CUFID (**C**omparative network analysis **U**sing the steady-state network **F**low to **ID**entify orthologous proteins) model assumes that if we pour water on the integrated network, there would be more water flow between the node pair as two nodes have high node similarity and they have a large number of neighboring nodes with a positive node similarity. Based on the assumption, we consider the water flow (or the steady-state network flow) as the node correspondence because the amount of water flow can be proportional to the node similarity as well as topological similarity. Then, to estimate the steady-state network flow over the integrated network, we design the specialized random walk protocol.

Given two biological networks G_X and G_Y, it constructs the integrated network by inserting pseudo edges between the node pairs $\left(u_i, v_j\right)$ if their node similarity is greater than a threshold. Then, it allows a single random walker to perform a random transition over the integrated network. If the random walker moves to its neighboring nodes through the original edges (i.e., protein–protein interactions) in the network G_X, the transition probability matrix can be obtained through the following equation:

$$\mathbf{P}_X = \mathbf{D}_X^{-1} \cdot \mathbf{A}_X, \tag{2.8}$$

where $\mathbf{A}_X$ is an adjacency matrix for the network G_X, and $\mathbf{D}_X$ is a diagonal matrix such that $D_X[i,i] = \sum_{\forall j} A_X[i,j]$. Similarly, if the random walker moves over the edges in the network G_Y, the transition probability matrix is given by

$$\mathbf{P}_Y = \mathbf{D}_Y^{-1} \cdot \mathbf{A}_Y. \tag{2.9}$$

Next, if the random walker changes its position through the pseudo edge connecting the nodes in the network G_X and the nodes in the network G_Y, the transition probability matrix is given by

$$\mathbf{P}_{X\to Y} = \mathbf{D}_S^{-1} \cdot \mathbf{S} \tag{2.10a}$$

and

$$\mathbf{P}_{Y\to X} = \mathbf{S}^{\mathrm{T}} \cdot \mathbf{D}_{S^{\mathrm{T}}}^{-1}, \tag{2.10b}$$

where $\mathbf{S}$ is $|N_X| \times |N_Y|$ dimensional node similarity matrix and $()^{\mathrm{T}}$ is a matrix transpose operation.

Finally, the transition probability matrix of the random walker over the integrated network can be obtained by concatenating the aforementioned transition matrices, where it is given by

$$\mathbf{P} = \begin{bmatrix} \mathbf{P}_X & \mathbf{P}_{X\to Y} \\ \mathbf{P}_{Y\to X} & \mathbf{P}_Y \end{bmatrix}. \tag{2.11}$$

Based on the transition probability matrix in Eq. (2.11), we can easily obtain the long-run proportion of time that the random walker spends at the particular node through the power method [19]. Although the SMRW and the CSRW models estimate the node correspondence through the long-run proportion of time that the random walkers spend at the node (or node pair), the CUFID model estimates the node correspondence through the staying time of the random walker at the pseudo edge connecting a node pair (u_i, v_j) in the integrated network, where it can be estimated as follows:

$$c(u_i, v_j) = \pi(u_i) \cdot P_{X\to Y}[v_j \mid u_i] + \pi(v_j) \cdot P_{Y\to X}[u_i \mid v_j], \tag{2.12}$$

where $\pi(u_x)$ is the steady-state probability of the random walker (i.e., the staying time of the random walker at the node u_x in the long-term point of view) and $P_{X\to Y}$ is the row-wise normalized node similarity score $\frac{s(u_i,v_j)}{\sum_{\forall v_j} s(u_i,v_j)}$ and $P_{Y\to X}$ is the column-wise normalized node similarity score $\frac{s(u_i,v_j)}{\sum_{\forall u_i} s(u_i,v_j)}$.

2.2.5 PPI Network Alignment Algorithms Using the Random Walk Models

Once we have estimated the node correspondence score through effective random walk models, it can be further enhanced through a probabilistic consistent transformation that can utilize the information from neighboring nodes in the pair of comparing networks [17]. The similar consistent transformation has been applied to obtain accurate biological sequence alignment [24].

The main idea of the probabilistic consistent transformation is based on the observation that proteins in a biological complex typically tend to be densely connected to each other and sparsely connected to the rest of the network. That is, if the network alignment algorithm can correctly align nodes in a conserved biological complex, their neighboring nodes also have higher chance to be correctly aligned. Hence, if the majority of neighboring nodes for u_i are well matched to the most of neighbors of v_j, it is reasonable that the alignment probability $P\left[u_i \sim v_j \mid G_X, G_Y\right]$ would be reinforced. Based on the background, the probabilistic consistent transformation is given by

$$\begin{aligned} P\left[u_i \sim v_j \mid G_X, G_Y\right] &= \alpha P\left[u_i \sim v_j \mid G_X, G_Y\right] \\ &\quad + (1-\alpha) \sum_{i'=1}^{|G_X|} \sum_{j'=1}^{|G_Y|} \left[P\left(u_i \sim u_{i'} \mid G_X\right) \right. \\ &\quad \left. P\left(u_{i'} \sim v_{j'} \mid G_X, G_Y\right) P\left(v_j \sim v_{j'} \mid G_Y\right) \right], \end{aligned} \tag{2.13}$$

where α is a weighting parameter to balance between the original alignment probability and the influence of the neighboring nodes.

Since the alignment probability can be represented a matrix, the above equation can be effective written in a matrix form as follows:

$$\mathbf{P}' = \alpha \mathbf{P} + (1-\alpha)\, \mathbf{G_X} \mathbf{P} \mathbf{G_Y}, \tag{2.14}$$

where $\mathbf{G_X}$ and $\mathbf{G_Y}$ is a stochastic matrix for the network G_X and G_Y, and $\mathbf{P}$ is a $|N_X| \times |N_Y|$ dimensional matrix representing the node correspondence.

Through the estimated and refined alignment probability, we can construct two different types of network alignments: (1) one-to-one mapping and (2) many-to-many mapping. The one-to-one mapping yields the best node-level mapping across different biological networks, where it is effective to predict the protein functional annotation based on the well-known protein functions. The many-to-many mapping identifies the group of conserved proteins in different biological networks, where it can be employed to understand biological mechanism as the proteins typically play together toward particular biological functions. The SMRW and CSRW models have been utilized in developing many-to-many node mapping based on the greedy algorithm [17, 18], and CUFID has been employed to develop one-to-one node mapping through the maximum weighted bipartite matching method [19].

2.3 Results

2.3.1 Datasets

We assessed the performance of global network alignment algorithms through synthetic PPI networks and real PPI networks. We utilized the synthetic networks provided in NAPAbench [30], where it provides a set of benchmark networks to assess the performance of pairwise, 5-way, and 8-way network alignment algorithms, and each dataset provides 10 network families that are created through different network growth models: CG (crystal growth), DMC (duplication–mutation–complementation), and DMR (duplication with random mutation). In this chapter, we utilized the pairwise network dataset that includes 3000 and 4000 nodes. The set of real PPI networks is obtained from IsoBase [31], where it provides comprehensive PPI networks for five model species by integrating several databases: BioGRID [27], DIP [28], HPRD [29], MINT [30], and IntAct [31]. The number of nodes and edges for five model species is summarized in Table 2.1. Note that we excluded the PPI network obtained from mouse because the number of edges is extremely smaller than other networks and we considered that it does not have a sufficient topological information.

2.3.2 Performance Metrics

We evaluated the quality of network alignments through two different aspects: biological significance and topological quality of alignments. The biological significance of the network alignment can be assessed through correct nodes (CN), specificity (SPE), mean normalized entropy (MNE), and gene ontology consistency (GOC) scores. The topological quality of the network alignment can be assessed through conserved interaction (CI), conserved orthologous interaction (COI).

Table 2.1 Number of nodes and edges for five PPI networks in IsoBase

Species	# nodes	# edges
H. sapiens (Human)	22,369	43,757
M. musculus (Mouse)	24,855	452
D. melanogaster (Fly)	14,098	26,726
C. elegans (Worm)	19,756	5853
S. cerevisiae (Yeast)	6659	38,109

First, given the network alignment results, we defined the set of the aligned nodes as the equivalence class. If the functional annotation for all nodes in the equivalence class has the consistent label, then the equivalence class is classified as the correct equivalence class. Next, CN is defined as the total number of nodes in the correct equivalence classes, and SPE is the ratio of the correct equivalence classes to the overall number of predicted equivalence classes. Note that for the real PPI networks, KEGG Orthology (KO) group annotations [32] can be utilized as functional annotations for each protein. Given an equivalence class C, the normalized entropy is given by $H(C) = -\frac{1}{\log d}\sum_{i=1}^{d} p_i \cdot \log p_i$, where p_i is the fraction of nodes in C having a functional label i and d is the total number of functional labels in C. Given the network alignment A, the GOC scores is given by $GOC(A) = \sum_{\forall (u_i \sim v_j) \in A} \frac{|GO(u_i) \cap GO(v_j)|}{|GO(u_i) \cup GO(v_j)|}$, where $GO(x)$ is the set of all GO terms [33] for the protein x. If the network alignment results the functionally consistent node mapping, it yields lower MNE and higher GOC scores. The topological quality of network alignment can be accessed through CI and COI. CI can be defined as the overall number of interactions between the equivalence classes, and COI is defined as the total number of interactions between the correct equivalence classes.

2.3.3 *Performance Assessment Through Synthetic PPI Networks*

We compared the performance of various network alignment algorithms: CSRW [18], SMETANA [17], BEAMS [10], PINALOG [21], and IsoRankN [3]. Note that SMETANA and CSRW are the network alignment algorithms using the SMRW and CSRW models, respectively. For PINALOG and SMETANA, we utilized the default parameters. For IsoRankN, we set the hyperparameter α to 0.6 as recommended in [3]. For BEAMS, we set the filtering threshold to 0.2 and α to 0.5 as shown in [10].

Table 2.2 shows that SMETANA and CSRW achieved higher CN and SPE compared to other algorithms. It indicates that the network alignment algorithm based

Table 2.2 Performance comparison for pairwise network alignment through synthetic networks in NAPAbench. Note that the bold face indicates the best performers

	DMC			DMR			CG		
	CN	SPE	MNE	CN	SPE	MNE	CN	SPE	MNE
CSRW	**5593.9**	**0.958**	**0.039**	**5305.3**	**0.939**	**0.055**	4893.2	0.942	0.054
SMETANA	5164.5	0.926	0.068	4900.6	0.916	0.078	4846.2	**0.949**	**0.048**
BEAMS	5076.5	0.826	0.150	5176.7	0.840	0.138	**5441.2**	0.870	0.112
PINALOG	3779	0.726	0.274	3533.4	0.683	0.317	4325	0.788	0.212
IsoRankN	3816.5	0.827	0.163	3905.2	0.836	0.155	3863.2	0.832	0.159

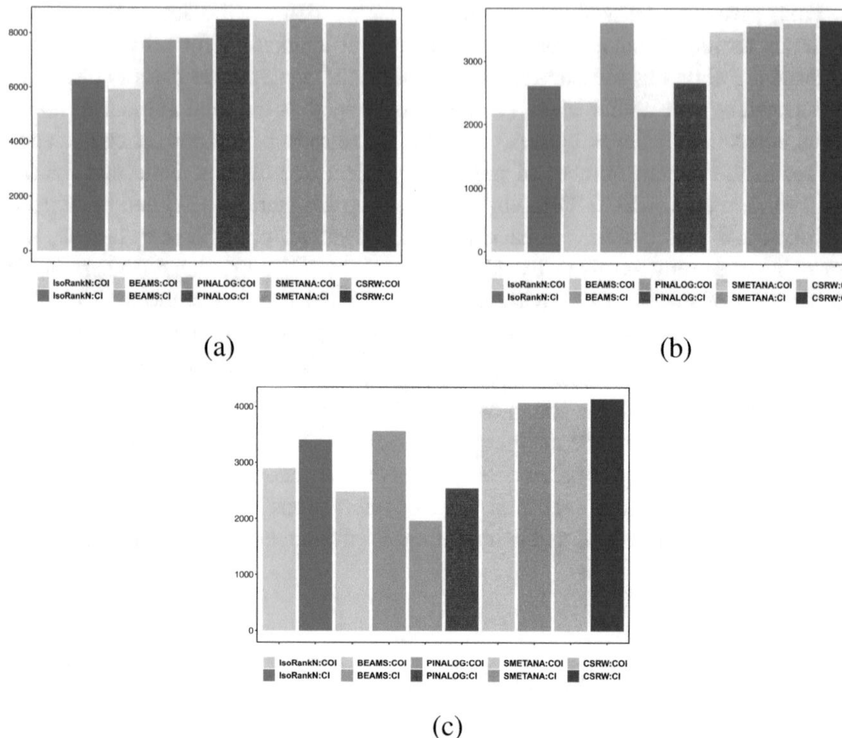

Fig. 2.2 The number of conserved orthologous interactions and conserved interactions for NAPAbench pairwise datasets. (**a**) CG. (**b**) DMC. (**c**) DMR

on the effective random walk models can predict more number of conserved nodes having the same biological function (i.e., CN) with high specificity and sensitivity. Although BEAMS can identify comparable number of correct nodes, its SPE is clearly lower than the random walk-based methods such as SMETANA and CSRW, where it means that BEAMS can yield considerable portion of equivalence classes including nodes with inconsistent functional labels. Additionally, the SMETANA and CSRW achieve remarkably lower MNE than other algorithms and it implies that the random walk-based algorithms can yield functionally coherent alignments compared to other algorithms.

Figure 2.2 shows the conserved interactions and conserved orthologous interactions for different network growth models in NAPAbench. It shows that SMETANA and CSRW can predict more number of conserved interactions for different network growth models. Since protein interactions typically activate or regulate biological mechanism, identifying larger number of conserved interactions is critical for comparative study. Additionally, most of identified conserved interactions are orthologous interactions, where it supports that the random walk-based comparative network analysis is very effective to deciphering and understanding conserved core biological mechanisms across different species. One possible explanation for achieving higher CI and COI is that the random walk models can effectively integrate the topological similarity and node-level similarity even though there are topological variations such as inserted or deleted nodes.

2.3.4 Performance Assessment Through Real PPI Network

We assessed the performance of various network alignment algorithms using real PPI networks in IsoBase [31]. In this experiment, we compared the performance of one-to-one network alignment algorithms and utilized KEGG Orthology (KO) group annotations to determine protein functions. That is, if the aligned proteins have the same KO group annotation, we considered that the proteins are correctly aligned. Note that, for a one-to-one pairwise network alignment, we omitted the mean normalized entropy in Table 2.3 because mean normalized entropy can be easily obtained through SPE (i.e., MNE = 1-SPE).

Table 2.3 shows the correct nodes and specificity for each pairwise network alignment result. As shown in Table 2.3, the network alignment algorithms based on the proposed random walk models (i.e., CUFID, CSRW, and SMETANA) can predict more number of correct nodes than other algorithms. For example, compared to other algorithms, there is a remarkable performance gap for the network alignment of fly and worm. Since one of the potential applications for one-to-one network alignment algorithm is predicting the protein functions in the less-studied species based on the functions of matching proteins in the well-studied species, network alignment algorithm that can identify more number of CN would

Table 2.3 Performance comparison for pairwise network alignment through IsoBase. Note that the bold face indicates the best performers

	Yeast–Human		Yeast–Fly		Fly–Human	
	CN	SPE	CN	SPE	CN	SPE
CUFID	**1330**	**0.736**	**1708**	0.748	**2528**	0.754
CSRW	1224	0.733	1610	**0.757**	2358	**0.763**
SMETANA	1134	0.71	1530	0.733	2096	0.706
PINALOG	1100	0.682	1368	0.722	1172	0.604
HubAlign	1082	0.633	1326	0.681	354	0.219
IsoRank	1142	0.702	1414	0.712	1736	0.725

	Yeast–Worm		Human–Worm		Fly–Worm	
	CN	SPE	CN	SPE	CN	SPE
CUFID	**1548**	0.834	**1858**	0.791	**2616**	**0.873**
CSRW	1426	**0.850**	1722	**0.803**	2444	0.870
SMETANA	1422	0.843	1570	0.780	2338	0.852
PINALOG	640	0.737	482	0.677	672	0.689
HubAlign	98	0.17	32	0.063	102	0.201
IsoRank	650	0.703	644	0.793	918	0.818

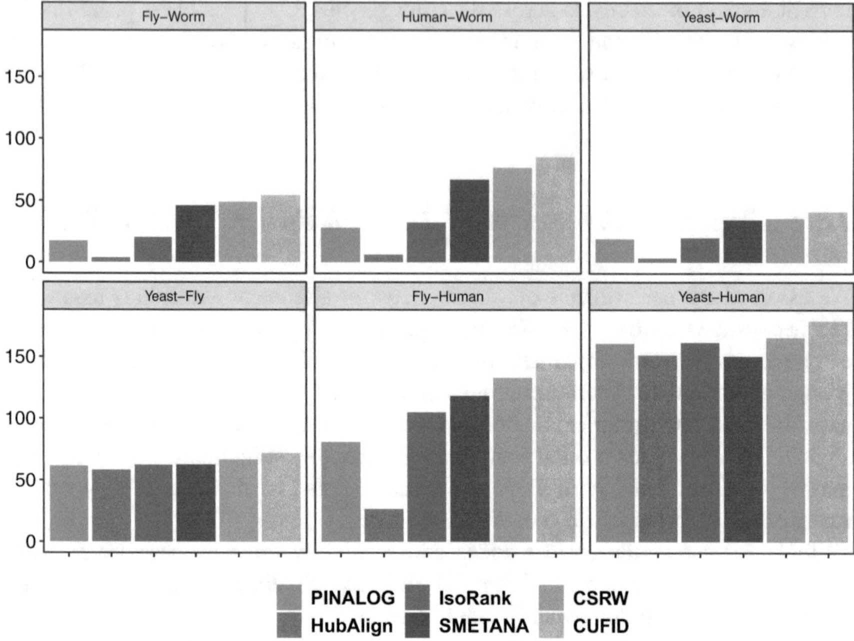

Fig. 2.3 The GOC scores for different network alignment results. The higher GOC scores indicate the higher functional consistency of aligned proteins

be much preferable. Additionally, the random walk-based methods achieved clearly higher SPE than other algorithms, where it supports the reliability of the random walk models to estimate the node correspondence. Next, as shown in Fig. 2.3, the random walk-based algorithms ranked in the best or next best performers for GOC scores. That is, performance assessment through multiple protein functional labels also shows that the aligned nodes obtained through the random walk models share more number of common GO terms than other algorithms, where it is a strong evidence of effectiveness in identifying and comparing the similarities and differences between different biological networks.

2.4 Discussion

Comparative network analysis provides an indispensable computational means to understand and decipher complex biological mechanisms across multiple species, and accurate estimation of node correspondence is a pivotal factor in developing reliable network alignment algorithms as the complexity and scale of biological data increase. This chapter introduces effective random walk models to estimate node correspondence across multiple biological networks by integrating topological similarity and node-level similarity. Additionally, through extensive performance evaluation using synthetic and real PPI networks, we verify that the accurate estimation of node correspondence using random walk models can help developing effective network alignment algorithms.

Through a comprehensive evaluation using synthetic and real PPI networks, we show the distinctive advantages of random walk models in developing the comparative network analysis algorithms. First, since the network can be effectively represented as the matrix, it can take advantage of various mathematical theories such as stochastic systems, graph theory, and linear algebra. Additionally, there are a number of effective and fast computational implementations and libraries for matrix operation so that random walk-based approach can be easily implemented as software packages. Next, random walk-based methods provide an intuitive model to integrate heterogeneous information such as topological similarity and node-level similarity, where it is crucial to estimate node correspondence. That is, random walk models are very flexible to deal with the topological variations so that it can lead to effective integration of heterogeneous information in a balanced manner. Finally, as the number of quality biological networks increases, there are increasing demands for the multiple network comparison tools. One discriminative advantage of random walk-based approach is that it can be easily expanded to deal with multiple networks. Although there are several advantages on random walk models in developing comparative network analysis methods, it has also limitations that should be addressed to derive sophisticated network analysis algorithms. For example, network representation using a matrix typically requires a huge amount of memory space because the number of nodes in the latest biological network typically spans up to tens of thousands. However, the nodes in biological networks generally have sparse interactions, and effective data structure such as sparse matrix representation can dramatically decrease the required computational resources.

Next, there is an important remark in developing network alignment algorithms. Due to the recent advances in high-throughput profiling techniques, protein–protein interactions have been archived in a number of public databases. Although each database has a PPI network for the same species, the amount of information archived in each database would not be identical. That is, a given species, the number of proteins and interactions could be different. As a result, performance assessments through a single PPI database would have a potential risk to developing a comparative network analysis tool that is overfitting in a particular PPI network. Hence, cross validation and performance evaluation across diverse PPI database could yield a balanced and robust network alignment algorithm (see Note 2 for the selected public PPI databases). Additionally, it would be highly recommended to assess various aspects of network alignment algorithms through a synthetic network of families [30, 34].

Finally, there is still room for performance improvement in comparative network analysis algorithms. For instance, novel machine learning techniques such as deep neural network can pave a new direction for comparative network analysis research. One of the major hurdles to employing cutting-edge machine learning techniques is the lack of effective vector representation of nodes in biological networks. However,

Table 2.4 List of selected public protein–protein interactions databases

Databases	# proteins	# interactions	URL
BioGRID	77,087	1,369,855	https://wiki.thebiogrid.org/
DIP	28,850	81,923	https://dip.doe-mbi.ucla.edu/dip/Stat.cgi
HPRD	30,047	41,327	https://hprd.org/
IntAct	115,379	1,035,669	https://www.ebi.ac.uk/intact/
IsoBase	87,737	114,897	http://cb.csail.mit.edu/cb/mna/isobase/#
MINT	26,344	131,695	https://mint.bio.uniroma2.it/
STRING	24,584,628	52,857,362	https://string-db.org/

there are several effective methods that can transform the nodes in a network into a low-dimensional vector by taking a topological structure into account [35, 36]. Combining a network representation learning and deep learning would have a great potential that can yield improved comparative network analysis results.

Notes

1. Biological sequence alignment algorithm aims to derive symbol (i.e., DNA, RNA, or amino acids) mapping between different biological sequences. Although there are matched symbols in different biological sequences, one symbol in a biological sequence cannot be mapped to another biological sequence because there are inserted or deleted symbols. The inserted symbol in a biological sequence can be considered to be mapped to the gaped symbol in another biological sequence. Similarly, although biological network alignments identify node mapping between different biological networks, there are inserted or deleted nodes that are not mapped to each other.
2. Table 2.4 shows the selected public PPI databases. As shown in Table 2.4, since the coverage of different PPI database, careful integration or refinement of raw PPI networks would be required to yield a quality test datasets.

References

1. Yoon, B.-J., Qian, X., Sahraeian, S.M.E.: Comparative analysis of biological networks: hidden Markov model and Markov chain-based approach. IEEE Signal Process. Mag. **29**(1), 22–34 (2011)
2. Singh, R., Xu, J., Berger, B.: Global alignment of multiple protein interaction networks with application to functional orthology detection. Proc. Natl. Acad. Sci. **105**(35), 12763–12768 (2008)
3. Liao, C.-S., Lu, K., Baym, M., Singh, R., Berger, B.: IsoRankN: spectral methods for global alignment of multiple protein networks. Bioinformatics **25**(12), i253–i258 (2009)

4. Andersen, R., Chung, F., Lang, K.: Local graph partitioning using pagerank vectors. In: 2006 47th Annual IEEE Symposium on Foundations of Computer Science (FOCS'06), pp. 475–486. IEEE, Piscataway (2006)
5. Kuchaiev, O., Milenković, T., Memišević, V., Hayes, W., Pržulj, N.: Topological network alignment uncovers biological function and phylogeny. J. R. Soc. Interface **7**(50), 1341–1354 (2010)
6. Kuchaiev, O., Pržulj, N.: Integrative network alignment reveals large regions of global network similarity in yeast and human. Bioinformatics **27**(10), 1390–1396 (2011)
7. Malod-Dognin, N., Pržulj, N.: L-GRAAL: Lagrangian graphlet-based network aligner. Bioinformatics **31**(13), 2182–2189 (2015)
8. Memišević, V., Pržulj, N.: C-GRAAL: common-neighbors-based global graph alignment of biological networks. Integr. Biol. **4**(7), 734–743 (2012)
9. Patro, R., Kingsford, C.: Global network alignment using multiscale spectral signatures. Bioinformatics **28**(23), 3105–3114 (2012)
10. Alkan, F., Erten, C.: Beams: backbone extraction and merge strategy for the global many-to-many alignment of multiple PPI networks. Bioinformatics **30**(4), 531–539 (2013)
11. Gligorijević, V., Malod-Dognin, N., Pržulj, N.: Fuse: multiple network alignment via data fusion. Bioinformatics **32**(8), 1195–1203 (2016)
12. Hu, J., Kehr, B., Reinert, K.: Netcoffee: a fast and accurate global alignment approach to identify functionally conserved proteins in multiple networks. Bioinformatics **30**(4), 540–548 (2014)
13. Hashemifar, S., Xu, J.: Hubalign: an accurate and efficient method for global alignment of protein–protein interaction networks. Bioinformatics **30**(17), i438–i444 (2014)
14. Mamano, N., Hayes, W.B.: Sana: simulated annealing far outperforms many other search algorithms for biological network alignment. Bioinformatics **33**(14), 2156–2164 (2017)
15. Saraph, V., Milenković, T.: Magna: maximizing accuracy in global network alignment. Bioinformatics **30**(20), 2931–2940 (2014)
16. Vijayan, V., Saraph, V., Milenković, T.: Magna++: maximizing accuracy in global network alignment via both node and edge conservation. Bioinformatics **31**(14), 2409–2411 (2015)
17. Sahraeian, S.M.E., Yoon, B.-J.: Smetana: accurate and scalable algorithm for probabilistic alignment of large-scale biological networks. PLoS One **8**(7), e67995 (2013)
18. Jeong, H., Yoon, B.-J.: Accurate multiple network alignment through context-sensitive random walk. In: BMC Systems Biology, vol. 9, p. S7. BioMed Central, London (2015)
19. Jeong, H., Qian, X., Yoon, B.-J.: Effective comparative analysis of protein-protein interaction networks by measuring the steady-state network flow using a Markov model. In: BMC Bioinformatics, vol. 17, p. 395. BioMed Central, London (2016)
20. Page, L., Brin, S., Motwani, R., Winograd, T.: The pagerank citation ranking: bringing order to the web. Technical report, Stanford InfoLab (1999)
21. Phan, H.T.T., Sternberg, M.J.E.: PINALOG: a novel approach to align protein interaction networks—implications for complex detection and function prediction. Bioinformatics **28**(9), 1239–1245 (2012)
22. Do, C.B., Mahabhashyam, M.S.P., Brudno, M., Batzoglou, S.: ProbCons: probabilistic consistency-based multiple sequence alignment. Genome Res. **15**(2), 330–340 (2005)
23. Roshan, U., Livesay, D.R.: Probalign: multiple sequence alignment using partition function posterior probabilities. Bioinformatics **22**(22), 2715–2721 (2006)
24. Sahraeian, S.M.E., Yoon, B.-J.: PicXAA: greedy probabilistic construction of maximum expected accuracy alignment of multiple sequences. Nucleic Acids Res. **38**(15), 4917–4928 (2010)
25. Sahraeian, S.M.E., Yoon, B.-J.: PicXAA-R: efficient structural alignment of multiple RNA sequences using a greedy approach. BMC Bioinf. **12**(1), S38 (2011)
26. Sahraeian, S.M.E., Yoon, B.-J.: PicXAA-Web: a web-based platform for non-progressive maximum expected accuracy alignment of multiple biological sequences. Nucleic Acids Res. **39**(suppl_2), W8–W12 (2011)

27. Sharan, R., Suthram, S., Kelley, R.M., Kuhn, T., McCuine, S., Uetz, P., Sittler, T., Karp, R.M., Ideker, T.: Conserved patterns of protein interaction in multiple species. Proc. Natl. Acad. Sci. **102**(6), 1974–1979 (2005)
28. Sharan, R., Ideker, T.: Modeling cellular machinery through biological network comparison. Nat. Biotechnol. **24**(4), 427 (2006)
29. Durbin, R., Eddy, S.R., Krogh, A., Mitchison, G.: Biological Sequence Analysis: Probabilistic Models of Proteins and Nucleic Acids. Cambridge University Press, Cambridge (1998)
30. Sahraeian, S.M.E., Yoon, B.-J.: A network synthesis model for generating protein interaction network families. PLoS One **7**(8), e41474 (2012)
31. Park, D., Singh, R., Baym, M., Liao, C.-S., Berger, B.: Isobase: a database of functionally related proteins across PPI networks. Nucleic Acids Res. **39**(suppl_1), D295–D300 (2010)
32. Ogata, H., Goto, S., Sato, K., Fujibuchi, W., Bono, H., Kanehisa, M.: Kegg: Kyoto encyclopedia of genes and genomes. Nucleic Acids Res. **27**(1), 29–34 (1999)
33. Ashburner, M., Ball, C.A., Blake, J.A., Botstein, D., Butler, H., Cherry, J.M., Davis, A.P., Dolinski, K., Dwight, S.S., Eppig, J.T., et al.: Gene ontology: tool for the unification of biology. Nat. Genet. **25**(1), 25–29 (2000)
34. Woo, H.-M., Jeong, H., Yoon, B.-J.: Napabench 2: a network synthesis algorithm for generating realistic protein-protein interaction (PPI) network families. PLoS One **15**(1), e0227598 (2020)
35. Perozzi, B., Al-Rfou, R., Skiena, S.: Deepwalk: online learning of social representations. In: Proceedings of the 20th ACM SIGKDD International Conference on Knowledge Discovery and Data Mining, pp. 701–710 (2014)
36. Grover, A., Leskovec, J.: node2vec: scalable feature learning for networks. In: Proceedings of the 22nd ACM SIGKDD International Conference on Knowledge Discovery and Data Mining, pp. 855–864 (2016)

Chapter 3
Computational Methods for Protein–Protein Interaction Network Alignment

Ruiquan Ge, Qing Wu, and Jinbo Xu

Abstract Advanced experimental and computational techniques have produced a large amount of biological data. Efficient analysis and mining of this complex data is critical to the understanding of biomechanism and evolution. Protein–protein interaction (PPI) data is important for understanding biological processes at the system level. Comparative analysis of PPI networks of various species may yield valuable information, such as conserved subnetwork motifs and pathways, and help with protein function prediction. Nevertheless, computationally PPI network alignment is a challenging problem and quite a few methods have been developed to address it.

This chapter presents three network alignment methods developed in the past few years based upon different principles. They combine sequence information, network topology, and subnetwork module information to score a PPI network alignment, and then employ two efficient algorithms (heuristics and convex optimization) to build alignments by optimizing the alignment scores. Experimental results show that these methods may build pairwise or multiple PPI network alignments efficiently, with accuracy favorably comparable to other methods.

Keywords Protein–protein interaction (PPI) network · Network alignment · Convex optimization · Network biology

3.1 General

It has been observed that proteins often bind together to form complexes/pathways to carry out specific functions. Protein–protein interaction (PPI) networks are important data for understanding biological dynamic processes such as metabolic

---

R. Ge · Q. Wu
Hangzhou Dianzi University, Hangzhou, China

J. Xu (✉)
Toyota Technological Institute at Chicago, Chicago, IL, USA

© Springer Nature Switzerland AG 2021
B.-J. Yoon, X. Qian (eds.), *Recent Advances in Biological Network Analysis*,
https://doi.org/10.1007/978-3-030-57173-3_3

pathways, DNA transcription and translation, and signaling cascades. Various experimental techniques have produced lots of PPI data for a number of species such as Yeast [1], Drosophila [2], and Helicobacter-pylori [3]. These techniques are often time-consuming and laborious and the resultant PPI data is usually noisy. Computational methods have also been developed to predict PPIs [4–7], which is complementary to the experimental techniques [8].

Comparative analysis and alignment of PPI networks may facilitate detection of functionally conserved pathways, transferring functional annotation among species, and predicting phylogenetic relationships of species. However, computationally PPI network alignment is a challenging problem because (1) it is hard to come up with an accurate scoring function to tell which alignments are the desirable; and (2) finding the optimal alignment of two or multiple networks is an NP-hard problem. In this chapter, we present three methods for PPI network alignment [9–11], which were developed in the past few years based upon different principles. These methods align two or multiple PPI networks by making use of protein sequence information, network topology, and network module decomposition. To build alignments efficiently, they employ both heuristics and convex optimization to find suboptimal alignments between two or multiple PPI networks.

3.1.1 Protein–Protein Interaction Network

A PPI network can be described by a graph $G = (V, E)$, where V is the set of nodes (proteins) and E is the set of edges (interactions). Let $N(u)$ denote the set of neighbors (i.e., interacting proteins) of a protein $u \in V$ and $|N(u)|$ the size of $N(u)$. $|N(u)|$ is also the degree of node u, denoted as $deg(u)$. Each edge in this graph represents one specific protein–protein interaction (PPI). For example, edge $e = (u, v)$ indicates that there is an interaction between two proteins u *and* v. Using some clustering methods, a PPI network may also be reorganized into a hierarchical binary tree structure [12], in which a leaf represents one protein and an internal tree node represents a subset (or group) of related proteins. Proteins in the same group are more densely connected and more likely to preform similar functions. PPI data may be generated by high-throughput experimental techniques [8] or computational methods [13, 14], which predict PPIs from information such as primary sequence, structure, expression, functional association, evolutionary conservation, and so on.

3.1.2 Protein–Protein Interaction Network Alignment

Roughly speaking, network alignment is to find common subgraphs across the input networks. Given two networks $G_1 = (V_1, E_1)$ and $G_2 = (V_2, E_2)$, their network alignment is a mapping between V_1 and V_2. There are two types of mappings in network alignment: one-to-one and many-to-many mappings. One-to-one mapping requires that a protein in one network be mapped to at most a protein in another. Many-to-many mapping allows that multiple proteins in one network may be mapped to multiple proteins in another [6].

Pairwise network alignment aligns proteins of only two networks. Along with more and more PPI networks are available, multiple network alignment (MNA) is also needed, which may offer more accurate alignments than pairwise alignment methods. Alignment of multiple PPI networks may help identify some subtly similar functional modules across multiple species.

Local and Global Network Alignment PPI networks can be aligned either locally or globally. Local network alignment (LNA) methods such as NetworkBlast [15] and AlignNemo [16] aim to detect conserved subnetworks, i.e., small isomorphic subnetworks corresponding to pathways and/or protein complexes. LNA may establish a one-to-one or many-to-many mapping between proteins. By contrast, global network alignment (GNA) methods maximize the overall match among input networks. That is, given a scoring function, GNA aims to find a highest-scoring alignment among two or multiple networks [17]. Again, both one-to-one and many-to-many mappings are allowed in GNA.

This chapter mainly focuses on global network alignment (GNA). A number of GNA methods have been developed including pairwise alignment methods IsoRank [6, 18] (which possibly is the first global alignment method), MI-GRAAL [19], GHOST [20], PISwap [21], and NETAL [22] and multiple alignment methods, e.g., NetworkBlast, Graemlin2.0, SMETANA [23], NetCoffee [24], BEAMS [25], FUSE [26], and IsoRankN [27]. IsoRank aligns two PPI networks by exploiting whether their interacting partners can match well. Graemlin2.0 uses a hill-climbing algorithm together with topology and phylogeny information [28, 29]. Both MI-GRAAL [19] and GHOST [20] use a seed-and-extend strategy to build alignment. PISwap uses IsoRank to generate an alignment. It iteratively swaps the edges in an alignment until reaching an optimum [21]. MAGNA uses genetic algorithm to search for the best alignment [30]. NETAL aligns two proteins based on their interacting partners.

In addition to network topology, many network alignment methods also make use of sequence information to improve alignment accuracy since PPI data is usually noisy and sequence information may reduce the impact of noise. The relative importance of sequence information is adjustable and depends on the quality of PPI data.

3.1.3 Experimental Datasets

We tested the alignment methods using several different types of PPI data. The first type includes five PPI networks obtained from HINT [31] for five species: *H. sapiens* (human), *S. cerevisiae* (yeast), *D. melanogaster* (fly), *C. elegance* (worm), and *M. musculus* (mouse). Table 3.1 shows their sizes.

Table 3.1 The sizes of the five PPI networks for five species

Species	Number of proteins	Number of interactions
Human	9336	29,617
Yeast	5169	20,176
Fly	7493	25,674
Worm	4494	11,292
Mouse	1298	2749

We also tested our methods using the PPI networks of two bacterial species: *Campylobacter jejuni* and Escherichia coli, which had the most complete PPI networks among all kinds of bacteria. The PPI network for Bacterium *Campylobacter jejuni* has 1111 nodes and 2988 edges. *Escherichia coli* is a model organism for studying the fundamental cellular processes such as gene expression and signaling. The *Escherichia coli* PPI network has 1941 nodes and 3989 edges. Finally, we also used some synthetic PPI network families in NAPAbench [32] benchmark generated by three network models: crystal growth [33], duplication-mutation-complementation [7], and duplication with random mutation [34].

3.1.4 Evaluation of Alignment Quality

We may evaluate the quality of an alignment both functionally and topologically [9, 29, 30]. Functional evaluation is particularly important when we want to identify functional submodules and transfer functional annotations through network alignment. Gene Ontology (GO) and KEGG Orthology (KO) annotations are used to measure the functional consistency and conservation of pathways. GO terms include biological process (BP), molecular function (MF), and cellular component (CC). KO annotations integrate comprehensive pathway information in KEGG [35]. In this chapter, we define a group of aligned proteins as a cluster. We say that a cluster is annotated if at least two of its proteins have GO or KO annotations. An annotated cluster is consistent if its annotated proteins share at least one common annotation.

3.1.4.1 Functional Metrics

Average of Functional Similarity (AFS) AFS is calculated according to the semantic similarity of GO terms. We may use Schlicker's similarity, which is based on the Resnik ontological similarity, to measure semantic similarity [36]. Let $S_{cat}(u, v)$ denote the GO functional similarity of two aligned proteins u and v in category cat (i.e. BP, MF or CC). AFS of an alignment A in category *cat* is calculated as follows:

$$\mathrm{AFS}_{\mathrm{cat}}(A) = \frac{1}{K}\sum\nolimits_{l=1}^{k}\left(\frac{1}{\mid A_l \mid}\sum_{v_i, v_j \in A_l, i \neq j} S_{\mathrm{cat}}\left(v_i, v_j\right)\right)$$

Functional Consistency (FC) *FC* is defined as the fraction of aligned proteins sharing common GO terms. The larger the fraction, the more biologically meaningful the alignment is.

Conserved Orthologous Interaction (COI) It is calculated as the total number of interactions between all consistent clusters. *COI* may be a better measure than *conserved (or aligned) interactions* since COI considers only aligned interactions formed by orthologous proteins. An alignment with a larger *COI* more likely contains functionally conserved subnetworks.

Mean Normalized Entropy (MNE) It measures functional coherence in a cluster. The lower the MNE, the more coherent a cluster is. MNE is defined as follows:

$$\mathrm{MNE} = \frac{1}{\log d}\sum_{i=1}^{d} p_i \times \log p_i$$

where d is the number of different GO annotations in the cluster and p_i represents the fraction of proteins with annotation GO_i.

Precision Precision is defined as the fraction of proteins in consistent clusters among the proteins in annotated clusters.

Sensitivity/Recall Sensitivity/Recall is defined as the total number of proteins in consistent clusters divided by the total number of proteins with at least one annotation.

Specificity Specificity is defined as ratio of consistent clusters to annotated clusters.

3.1.4.2 Topological Metrics

The measures for the topological quality are as follows.

Symmetric Substructure Score (S^3) S^3 penalizes the alignment that map sparse regions of a network to denser ones and vice versa. Let $G[V]$ denote the induced subnetwork of G with node set V and $E(G)$ denote the edge set of G. Let $f(E_1) = \{(g(u), g(v)) \in E_2, u, v \in E_1\}$ and $f(V_1) = \{g(v) \in V_2, v \in V_1\}$. Mathematically, S^3 is defined as follows:

$$\mathrm{S}^3 = \frac{\mid f(E_1) \mid}{\mid E_1 \mid + \mid E(G_2[f(V_1)]) \mid + \mid f(E_1) \mid} \times 100$$

Largest Common Connected Subgraph (LCCS) It is calculated as the number of edges in the largest connected subgraph in an alignment. Larger and denser subgraphs give more insight into common topology of the network. In addition, the larger and denser subgraphs may be more biologically important, as Bader has shown that a dense PPI subnetwork may correspond to a vital protein complex [37]. S^3 and LCCS have been used only for evaluation of pairwise alignments.

Conserved Interaction (CI) It is also called edge correctness (EC), which is calculated as the ratio between the number of aligned interactions and the number of interactions.

c-Coverage It is the number of clusters composed of proteins from exactly c species. Specifically, total coverage is the number of clusters composed of proteins from at least two species. Clusters with a large c usually imply that proteins in these clusters are conserved across multiple species.

3.2 Alignment Methods

3.2.1 HubAlign: PPI Network Alignment by Topological Importance

3.2.1.1 Introduction

A biological network usually contains some topologically and functionally important proteins denoted as hubs and bottlenecks. Hub proteins play a central role in biological processes and may be involved in various functional modules. Bottlenecks refer to those proteins with a high betweenness centrality [38]. Hubs and bottlenecks tend to mutate slowly and are conserved, so they are more likely to be aligned. Removing the hub and bottleneck proteins may separate a PPI network and disrupt some cooperated functional modules. Since hub and bottleneck proteins are functionally important and conserved, they are more likely to be aligned. Therefore, HubAlign first aligns those functionally important and conserved proteins and then the less important proteins. HubAlign uses an iterative minimum-degree heuristic method to estimate the relative importance of one protein with respect to the whole network. Afterwards, it aligns two proteins using a greedy algorithm by combining topological importance score and sequence similarity. Experimental results show that HubAlign not only runs fast, but also compares favorably to popular methods such as IsoRank [6, 18], MI-GRAAL [19], GHOST [20], PISwap [21] in terms of alignment accuracy.

3.2.1.2 Method Details of HubAlign

Topological and Functional Importance of Proteins The relative importance of a protein or interaction (or edge) in a PPI network shows its role in maintaining network structure or function [39]. The degree of one protein reflects only the local property of a network [40] and the edge-betweenness only takes into consideration the shortest paths in a network. Instead of using degree or edge-betweenness, HubAlign calculates the topological importance of proteins and interactions using a method similar to graph tree-decomposition, which decomposes a graph into a tree of subnetwork modules. Each module is a highly connected subgraph and its topological complexity reflects interaction strength of proteins in this module. HubAlign calculates the topological importance using a minimum-degree heuristics method, taking into consideration only those proteins with degree no more than d where d is a small constant (e.g., 10) to be determined empirically. The value of d should not be very large in case that the deletion of the high-degree proteins (e.g., hubs) destroys the whole network functionally or structurally [39, 41]. Starting from the proteins with degree one, HubAlign assigns an initial weight to individual proteins and interactions as follows:

$$w(e) = \begin{cases} 1 & e \in E \\ 0 & \text{otherwise} \end{cases}, \; w(u) = 0 \quad \forall u \in V$$

where $w(e)$ and $w(u)$ represent the weight of interaction e and protein u, respectively. The interaction weight may also be initialized by the PPI confidence score if it is available in the PPI data. Then, HubAlign repeatedly removes the proteins with minimum degree and adjusts weight. When one protein is removed, its interactions are also removed and the weight of the removed proteins and interactions are allocated to their interacting proteins and related interactions as follows:

$$\begin{cases} \forall v \in N(u) : w(v) = w(v) + w(u) + w\,(u,\,v) & \deg(u) = 1 \\ \forall v_1,\,v_2 \in N(u) : w\,(v_1,\,v_2) = w\,(v_1,\,v_2) + \frac{w(u)+\sum_{v\in N(u)} w(u,v)}{\frac{|N(u)||N(u)-1|}{2}} & \deg(u) > 1 \end{cases}$$

Afterwards, an important score of one protein is calculated by combining both protein and interaction weight as follows:

$$S(v) = w(v) + \lambda \sum_{u \in V} w\,(u,\,v)$$

where $S(v)$ is the score of protein v and λ regulates the importance of the weight between interaction and individual proteins (empirically $\lambda = 0.1$). Finally, $S(v)$ is normalized as follows to reduce the impact of network size:

$$S(v) = \frac{S(v)}{\max_{v \in V}\{\mathrm{S(v)}\}}$$

Algorithm for Building the Alignment The normalized S score reflects the global topological importance of one protein in a PPI network. Two proteins with similar S scores are more likely to be aligned. The topological similarity between two proteins $u \in V_1$ and $v \in V_2$ is calculated as follows:

$$A\,(u,\,v) = \min\,(S(u),\,S(v))$$

HubAlign also allows the inclusion of sequence homology information (i.e., sequence similarity) into the alignment score. Let $B(u, v)$ denotes the normalized BLAST bitscore for two proteins u and v. The final alignment score between two proteins is defined as follows:

$$A^*\,(u,\,v) = \alpha \times A\,(u,\,v) + (1-\alpha) \times B\,(u,\,v)$$

where $0 \leq \alpha \leq 1$ is a tuning parameter that controls the contribution of sequence similarity and topological information. Sequence information may be used to reduce the impact of the noisy PPI data. When HubAlign was developed, the default value of α was set to 0.7. Nevertheless, along with more accurate PPI data being generated, a larger value may be used for α. Generally speaking, network topology information shall play a more important role when we want to annotate protein functions or identify functionally conserved network modules which cannot be fulfilled based upon sequence information.

After the alignment score of all protein pairs is calculated, HubAlign generates a global alignment of two PPI networks using a greedy algorithm. HubAlign first identifies the pair of proteins with the highest alignment score as a seed alignment and then greedily extends the alignment. After aligning a pair of nodes, HubAlign aligns their neighbors as long as their alignment scores are not bad. When HubAlign finishes extending alignment from an initial seed, it chooses a new seed from the remaining unaligned proteins and do alignment extension again. HubAlign repeats this procedure until all proteins in the smaller PPI network are aligned to the larger network.

Time Complexity Let $n = \max\{|V_1|, |V_2|\}$. At first, it takes $O(n)$ time to find the node with minimum degree. Since only the proteins with degree less than 10 are removed, it takes $O(1)$ to update the weight of the neighbors (i.e., interacting proteins). Further, at most n proteins are removed, so the total time complexity of assigning weight is $O(n^2)$. Afterwards, it takes $O(n^2)$ time to calculate the alignment score of all protein pairs between two PPI networks. At last, it takes $O(n^2)$ to select a seed protein pair. For alignment extension, HubAlign employs a priority queue to save the neighbors of each pair of aligned proteins. It takes $O(n\log(n))$ to update the priority queue and constant time to extract the pair with highest score from the queue. Therefore, aligning two networks of n proteins takes $O(n^2\log(n))$ time. Overall, the total time complexity of aligning two networks is $O(n^2\log(n))$.

HubAlign is much more efficient than many network alignment methods developed before it. Running on a Linux system with 1400 MHz CPU and 2GB RAM, it took NETAL [22], HubAlign, IsoRank [6, 18], MI-GRAAL [19], and GHOST [20] 80, 412, 7610, 78,525, and 3037 s, respectively, to align the yeast and human PPI networks.

3.2.1.3 Performance of HubAlign

We have compared HubAlign with several global network alignment methods including IsoRank, MI-GRAAL, NETAL, GHOST, and PISwap. Meanwhile, PISwap used the alignment produced by GRAAL and IsoRank as input. The yeast and human PPI networks taken from IntAct [42] were used to test these methods. The yeast PPI network has 5673 nodes and 49,830 edges, while the human PPI network consists of 9002 nodes and 34,935 edges. As shown in Table 3.2, HubAlign produces an alignment with much larger EC, LCCS, and S^3 than the other methods except NETAL. To measure FC and AFS of an alignment, all the aligned proteins pairs which have GO annotations are examined. Table 3.2 shows that HubAlign yields alignments with significantly higher AFS than the other methods, especially when biological process (BP) and molecular function (MF) are considered.

Our experimental results on the five PPI networks (shown in Table 3.1) indicate that HubAlign can align more functionally similar proteins and find larger complexes that are topologically or biologically significant. On the two bacterial species *Campylobacter jejuni* and Escherichia coli, HubAlign outperformed the other methods although all the AFS values are small due to insufficient GO annotations of the bacterial proteins in terms of AFS.

HubAlign has two hyperparameters λ and α. Meanwhile, λ determines the relative importance of interaction score and protein score, and α determines the relative importance of sequence and topology information. Testing on the yeast and human PPI networks with respect to different values of λ and α, our results indicate

Table 3.2 Quality of human-yeast alignments generated by various methods

Method	EC	LCCS	S^3	AFS(BP)	AFS(MF)	AFS(CC)
IsoRank	2.12	44	1.23	0.76	0.63	0.77
MI-GRAAL	13.87	4832	8.12	0.63	0.52	0.72
GHOST	17.04	7000	13.59	0.82	0.66	0.83
PISwap	2.16	62	1.23	0.77	0.63	0.77
NETAL	28.65	9695	20.16	0.58	0.46	0.71
HubAlign	21.59	7240	14.67	0.95	0.81	0.88

that when λ is in the range (0.1, 0.2), the resultant alignment has a good tradeoff between topological and biological quality. Similarly, when α is in the range (0.7, 1), the resultant alignment has a good tradeoff between topological and biological quality.

3.2.2 *ModuleAlign: Module-based Global Alignment of PPI Networks*

3.2.2.1 Introduction

ModuleAlign was developed with the goal of producing more accurate alignments than HubAlign [10]. Similar to HubAlign, ModuleAlign makes use of both protein sequence and network topology information [9]. However, ModuleAlign also makes use of information in subnetwork modules. In particular, ModuleAlign first decomposes a PPI network into a hierarchical tree of subnetwork modules and then calculates the similarity of two proteins in the context of network modules. Our experimental results showed that ModuleAlign produced better alignments than GHOST, MAGNA++, NETAL, HubAlign, and L-GRAAL.

3.2.2.2 Method Details of ModuleAlign

Module-based Alignment Score ModuleAlign first employs a tool HAC-ML [43] to detect subnetwork modules in a PPI network, organizing it as a tree of subnetwork modules. High-level modules (level > 3) in the tree are ignored since they may have too many proteins. Afterwards, ModuleAlign calculates the similarity between each pair of modules $cl \in C_1$ and $cl' \in C_2$, where C_1 and C_2 represent the set of modules of two input networks, respectively. The similarity between two modules *clusterSim*(cl, cl') is calculated as the average BLAST score of all protein pairs across these two modules as follows:

$$\text{clusterSim}\left(cl, cl'\right) = \frac{\sum_{u \in cl, v \in cl'} \text{blast}\,(u, v)}{\mid cl \mid * \mid cl' \mid}$$

where *blast*(u, v) is the BLAST score between proteins u and v. Then ModuleAlign calculates the homology score of two proteins using the module similarity score as follows:

$$\text{HS}\,(u, v) = \frac{\sum_{\substack{cl \in C_1 : u \in cl, \\ cl' \in C_2 : v \in cl'}} \text{clusterSim}\left(cl, cl'\right)}{\mid C_1 \mid * \mid C_2 \mid}$$

The alignment score employed by ModuleAlign consists of both homology score HS(u, v) and topology score TS(u, v). That is, the score of two aligned proteins is defined as follows:

$$A\left(u,\ v\right) = \alpha \times \mathrm{HS}\left(u,\ v\right) + (1-\alpha)\,\mathrm{TS}\left(u,\ v\right)$$

Meanwhile, TS(u, v) is the topology score used by HubAlign and HS(u, v) is specific to ModuleAlign;$0 \leq \alpha \leq 1$ is used to balance the contribution of topology score and homology score. In the implementation of ModuleAlign, α is set to 0.4 by default.

Alignment Strategy of ModuleAlign The alignment algorithm mainly consists of two steps. First, the Hungarian method [44] is applied to compute an initial alignment between two networks. This step may find pairs of proteins which may have similar topology properties and/or function. Second, ModuleAlign maximizes the number of evolutionarily conserved interactions. Based upon the initial alignment, ModuleAlign iteratively includes protein pairs with the best alignment score into the alignment. After one protein pair (u_b, v_b) is included into the alignment, for each neighbor u of $\mathrm{u_b}$, ModuleAlign removes its current alignment and adjust the alignment score between u and all the neighbors of v_b as follows:

$$\forall_{v \ni N(v_b)} A\left(u,\ v\right) = A\left(u,\ v\right) + 1/\max_{u' \ni V_1 \cup V_2}\left\{S\left(u'\right)\right\}$$

where $\max_{u' \ni V_1 \cup V_2}\left\{S\left(u'\right)\right\}$ is the same normalization factor used in the topology score. Then, ModuleAlign performs one primal-dual iteration of the Hungarian algorithm to update the alignment scores. This procedure will stop when all proteins of the smaller network are aligned.

Time Complexity Let $n = \max(|V_1|, |V_2|)$. It takes $O(n^3\log(n))$ time to decompose the input networks into modules and $O(n^2)$ to calculate the topology score. Calculating the initial alignment scores needs $O(n^3\log(n))$. In the second step, it takes $O(n^2)$ to update the alignment scores. Finally, it takes $O(n^3)$ to iteratively run the Hungarian method. In summary, ModuleAlign has time complexity $O(n^3\log(n))$. Empirically, on a Linux system with 1400 MHz CPU and 2GB RAM, it took NETAL, HubAlign, MAGNA++, L-GRAAL, GHOST, and ModuleAlign 4, 15, 621, 64, 41, and 15 min, respectively, to align the yeast and human networks, and 7, 25, 757, 79, 50, and 30 min, respectively, to align the fly and human networks.

3.2.2.3 Performance of ModuleAlign

We have tested ModuleAlign on the five PPI networks in Table 3.1 and compared it with several global network alignment methods including NETAL, GHOST, HubAlign, MAGAN++, and L-GRAAL, which represented the state of the art when ModuleAlign was developed. We ran the other methods (except MAGAN++) with the default parameters. For MAGAN++, we ran its genetic algorithm over

Table 3.3 Evaluation of the alignment on the bacterial PPI networks

Method	Precision	Recall	AFS(BP)	AFS(MF)	EC	S^3
NETAL	0	0	0.15	0.09	32.36	19.54
GHOST	0.05	0.04	0.19	0.14	22.79	15.14
HubAlign	0.31	0.24	0.25	0.22	24.65	16.51
MAGNA++	0	0	0.24	0.18	24.83	19.32
L-GRAAL	0.11	0.09	0.12	0.10	24.61	17.83
ModuleAlign	**0.37**	0.30	0.31	0.28	25.95	16.92

15,000 generations with a population size of 2000. The alignment quality is evaluated in terms of both functional and topological consistency. KEGG Orthology (KO) annotations are employed to calculate the functional consistency of an alignment because orthologous genes often share identical KO annotations. In terms of precision and recall, ModuleAlign significantly outperformed the other methods and had the highest accuracy in terms of predicting the consistent clusters, which are important for studying the orthologous relationship and evolutionary pathways between species. We measure AFS of an alignment using GO terms [45]. ModuleAlign generated alignment with better AFS in all three categories BP, MF, and CC. ModuleAlign not only generated more functionally consistent protein pairs, but also produced alignment with larger EC, LCCS, and S^3 values than the other methods except NETAL. NETAL had a higher EC, LCCS, and S^3 scores, but has a lower AFS. In the alignment generated by ModuleAlign for yeast and human, there are two conserved subnetworks which could not be identified by the other methods. Proteins in each of the two subnetworks have high functional coherence and very similar KO annotations.

We also tested ModuleAlign on the PPI networks of two bacterial species *Escherichia coli* (*E. coli*) and *Campylobacter jejuni* (*C. jejuni*) [46, 47]. As shown in Table 3.3, ModuleAlign has significantly higher precision and better AFS (BP, MF) than the other methods.

3.2.3 ConvexAlign: Multiple Network Alignment via Convex Optimization

3.2.3.1 Introduction

Before we have described two methods HubAlign and ModuleAlign for the global alignment of two PPI networks. Here we describe an algorithm ConvexAlign [11] for the global alignment of multiple PPI networks. Many multiple network alignment (MNA) methods are just a simple extension of pairwise alignment methods [23, 27]. That is, they first build pairwise alignments and then gradually merge them into a multiple alignment in a progressive manner. This progressive strategy is easy to implement and runs fast, but its major issue is that alignment errors generated at earlier stages can hardly be fixed later. Instead of using a progressive strategy, ConvexAlign align all input PPI networks simultaneously by formulating this problem as a convex optimization problem. Since ConvexAlign conducts simultaneous alignment of all networks, it greatly reduces alignment errors. ConvexAlign integrates network topology, sequence similarity, and interaction conservation to score an alignment and then uses ADMM [48] (alternating direction method of multipliers) to solve the convex optimization problem. Our experiments indicate that ConvexAlign outperformed state-of-the-art methods and detected a few conserved complexes which could not be detected by others.

3.2.3.2 Method Details of ConvexAlign

Alignment Score ConvexAlign uses a scoring function consisting of node and edge scores where nodes indicate proteins and edges represent interactions. The score of one alignment A is calculated as below:

$$f(A) = (1 - \lambda_2)\, f_{\text{node}}(A) + \lambda_2 f_{\text{edge}}(A)$$

where λ_2 is the weight factor; f_{node} and f_{edge} denote the node (protein) similarity and interaction conservation, respectively. $f_{\text{node}}(A)$ and $f_{\text{edge}}(A)$ sum the scores among all pairs of aligned proteins and all aligned interaction pairs, respectively.

$$f_{\text{node}}(A) = \sum_{1 \le i \le j \le N} \sum_{A_k \in A, v_i, v_j \in A_k} \text{node}\left(v_i, v_j\right)$$

$$f_{\text{edge}}(A) = \sum_{1 \le i \le j \le N} \sum_{\substack{A_k,\, A_l \in A, \\ v_i,\, v_j \in A_k, v'_i,\, v'_j, \in A_l}} \delta\left(\left(v_i, v'_i\right) \in E_i\right) \delta\left(\left(v_j, v'_j\right) \in E_j\right)$$

where $node(v_i, v_j)$ denotes the topology and sequence similarity score used by HubAlign; and $\delta\left(\left(v_i, v'_i\right) \in E_i\right)$ is an indicator function.

Let $M = \sum_{i=1}^{N} \mid V_i \mid$ denote the number of proteins in all the input PPI networks. We may represent a valid multiple alignment A by a binary matrix $Y = (Y_1; Y_2; \ldots; Y_N) \in \{0,1\}^{M \cdot K}$, where $Y_i\ (v_i, A_j) = 1$ if and only if v_i is in group A_j. Each group is a set of proteins aligned to each other. In addition, inspired by Huang and Guibas [49], we also introduce an alignment matrix X as follows:

$$X = \begin{pmatrix} I_{|V_1|} & X_{12} & \ldots & X_{1N} \\ X_{12}^T & I_{|V_2|} & \ldots & X_{2N} \\ \ldots & \ldots\ldots & & \ldots \\ X_{1N}^T & \ldots & \ldots & I_{|V_N|} \end{pmatrix} = \begin{pmatrix} Y_1 \\ Y_2 \\ \ldots \\ Y_N \end{pmatrix} \cdot \begin{pmatrix} Y_1 \\ Y_2 \\ \ldots \\ Y_N \end{pmatrix}^T$$

where X is a positive semidefinite matrix encoding a one-to-one global alignment. Using X, we may have a linear scoring function:

$$f_{\text{node}} = \sum_{1 \le i < j \le N} \sum_{v \in V_i, v' \in V_j} \text{node}\left(v, v'\right) X_{ij}\left(v, v'\right) = \sum_{1 \le i < j \le N} \left\langle C_{ij}, X_{ij} \right\rangle$$

$$f_{\text{edge}} = \sum_{1 \le i < j \le N} \sum_{\left(v_i, v'_i,\right) \in E_{i,} \left(v_j, v'_j,\right) \in E_{j,}} y_{ij}\left(v_i, v_j, v'_i, v'_j\right) = \sum_{1 \le i < j \le N} \left\langle \overrightarrow{1}, y_{ij} \right\rangle$$

where C_{ij} is a matrix composed of the values of node(v, v') and $y_{ij}\left(v_i, v_j, v'_i, v'_j\right) = X_{ij}\left(v_i, v_j\right) X_{ij}\left(v'_i, v'_j\right)$. The nonlinear constraint between y_{ij} and X_{ij} can be replaced by the linear inequalities as follows [50, 51]:

$$\forall v_i \in V_i, \quad \sum_{vj:\left(v_j, v'_j,\right) \in E_{j,}} y\left(v_i, v_j, v'_i, v'_j\right) \le X_{ij}\left(v'_i, v'_j\right)$$

$$\forall v_j \in V_j, \quad \sum_{v_i:\left(v_i, v'_i,\right) \in E_{i,}} y\left(v_i, v_j, v'_i, v'_j\right) \le X_{ij}\left(v'_i, v'_j\right)$$

$$\forall v'_i \in V_i, \quad \sum_{v'_j:\left(v_j, v'_j,\right) \in E_{j,}} y\left(v_i, v_j, v'_i, v'_j\right) \le X_{ij}\left(v_i, v_j\right)$$

$$\forall v'_j \in V_j, \quad \sum_{v'_i:\left(v_i, v'_i,\right) \in E_{i,}} y\left(v_i, v_j, v'_i, v'_j\right) \le X_{ij}\left(v_i, v_j\right)$$

The above formulas are equivalent to $y_{ij}\left(v_i, v_j, v'_i, v'_j\right) = X_{ij}\left(v_i, v_j\right) X_{ij}\left(v'_i, v'_j\right)$ when both y_{ij} and X_{ij} are binary variables and one-to-one mapping is enforced. We may also simplify the above formulas as follows:

$$B_{ij} y_{ij} \le \mathcal{F}_{ij}\left(X_{ij}\right)$$

where B_{ij} is the coefficient and $\mathcal{F}_{ij}\left(X_{ij}\left(v_i, v_j\right)\right) =< P_{ij}, X_{ij} >$. Only one element $P_{ij}(v_i, v_j)$ is equal to 1 in P_{ij}. $\mathcal{F}_{ij}$ is a linear operator that maps the corresponding element of X_{ij} for each constraint.

Finally, we have the following semidefinite program for one-to-one alignment of multiple networks:

$$\text{maximize} \sum_{1 \le i < j \le N} (1 - \lambda_2) \langle C_{ij}, X_{ij} \rangle + \lambda_2 \left\langle \vec{1}, y_{ij} \right\rangle$$

$$\text{subject to } y_{ij} \in \{0, 1\}^{|E_i| \times |E_j|}, B_{ij} y_{ij} \le \mathcal{F}_{ij} \left(X_{ij} \right), 1 \le i < j \le N$$

$$X_{ij} \vec{1} \le \vec{1}, X_{ij}^T \vec{1} \le \vec{1}, X_{ij} \in \{0, 1\}^{|V_i| \times |V_j|}, 1 \le i < j \le N$$

$$\mathrm{X} \succcurlyeq 0, X_{ii} = I_{|V_i|}, 1 \le i \le N.$$

After relaxing y_{ij} and X_{ij} to real values between 0 and 1, the convex program are as follows:

$$\text{maximize} \sum_{1 \le i < j \le N} (1 - \lambda_2) \langle C_{ij}, X_{ij} \rangle + \lambda_2 \left\langle \vec{1}, y_{ij} \right\rangle$$

$$\text{subject to } y_{ij} \ge 0, B_{ij} y_{ij} \le \mathcal{F}_{ij} \left(X_{ij} \right), 1 \le i < j \le N$$

$$X_{ij} \vec{1} \le \vec{1}, X_{ij}^T \vec{1} \le \vec{1}, X_{ij} \ge 0, 1 \le i < j \le N$$

$$X \succcurlyeq 0, X_{ii} = I_{|V_i|}, 1 \le i \le N.$$

The constraint $X \succcurlyeq 0$ enforces cycle consistency in the alignments.

We first employed ADMM to solve the convex optimization problem, and then a greedy rounding strategy to convert fractional solution to integral. Although conducting simultaneous alignment, ConvexAlign is computationally efficient. It took ConvexAlign, IsoRankN, BEAMS, FUSE, NetCoffee, and SMETANA 480, 1129, 900, 780, 15, and 37 min, respectively, to build the global alignment for the PPI networks of five species (human, yeast, fly, worm, and mouse) [31] described in Table 3.1. That is, ConvexAlign is only slower than NetCofee and SMETANA, but NetCoffee does not generate alignments with good functional consistency.

3.2.3.3 Performance of ConvexAlign

We have compared ConvexAlign with a few methods including BEAMS [25], FUSE [26], IsoRankN [27], NetCoffee [24], and SMETANA [23] on the 5 PPI networks in Table 3.1. We ran SMETANA, NetCofee, and FUSE with their default parameters, and BEAMS and IsoRankN with three different values for their parameter $\alpha = \{0.3, 0.5, 0.7\}$. We also tested ConvexAlign on the synthetic PPI networks in the NAPAbench benchmark described in Sect. 1.3. We used the eight-way alignment dataset, which contains 3 network families, each with 8 networks of 1000 proteins. The dataset simulates a network family of closely related species, so this benchmark has very different properties than the five real PPI networks in Table 3.1.

We evaluated the ConvexAlign alignments by several topological and functional consistency metrics described in Sect. 1.4. On the real data, ConvexAlign has a larger c-coverage ($c = 4, 5$) than the other methods except SMETANA and NetCoffee. ConvexAlign has a better CI than all other methods except SMETANA. These conserved interactions can help to identify the functional modules conserved among networks of different species. ConvexAlign has much higher specificity than the others when all the annotated clusters ($c \geq 2$) are considered. Experimental results suggest that ConvexAlign finds more functionally consistent clusters. These clusters may be used to analyze the orthologous relationship of the proteins, identify the conserved subnetworks, and predict the function of unannotated proteins. ConvexAlign also outperforms other methods in terms of MNE and sensitivity. The average of functional similarity ($\overline{\text{AFS}}$) of ConvexAlign is 6–20% larger than the other methods for clusters composed of proteins in three, four, and five species. The distribution of AFS scores further confirms that ConvexAlign yields clusters with higher functional similarity in terms of MF and BP. On the synthetic data, ConvexAlign has a better specificity than the other methods when $c = 4, 5, 6, 7, 8$ species are considered. FUSE only finds none or a few (1 or 2) consistent clusters when $c = 3, 4, 5, 6$. In terms of the number of consistent clusters, ConvexAlign is slightly better than the second best method BEAMS regardless of c and much better than the others. In terms of MNE, COI, and RCI, our experiments also show that ConvexAlign outperforms all the other methods.

One of the applications of PPI network alignment is to identify conserved subnetworks in the species which cannot be found by sequence similarity alone. Figure 3.1 shows one conserved complex detected by ConvexAlign related to DNA replication (with p-value $<6^{-10}$), which could not be detected by other methods.

3.3 Conclusion

Protein–protein interaction (PPI) networks are an effective tool for modeling and understanding complex organization of biological processes at the system level. Comparative analysis and alignment of PPI networks may yield useful information to facilitate identification of functionally conserved interactions and network mod-

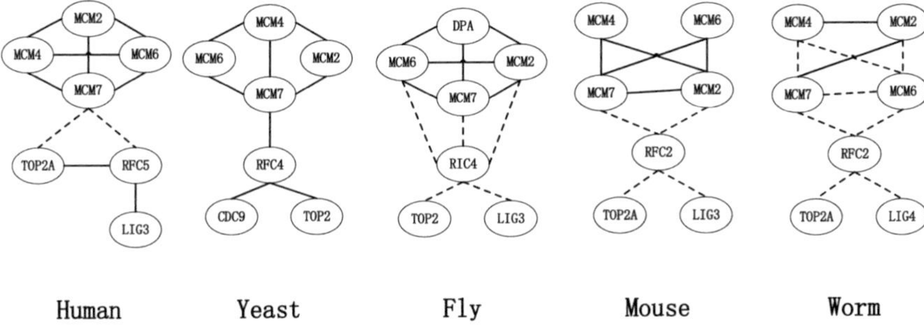

Fig. 3.1 The DNA replication complex in each input PPI network detected by ConvexAlign

ules and even disease-related genes. This chapter describes three methods for PPI network alignment: HubAlign and ModuleAlign for pairwise alignment and ConvexAlign for multiple alignment. HubAlign makes use of topological importance, ModuleAlign makes use of the network modules in addition to topological importance, and ConvexAlign aligns multiple PPI networks simultaneously. Compared to other methods, these three methods are not only computationally efficient, but also have favorable accuracy. They have also identified some functionally conversed subnetwork modules which cannot be detected by other methods.

References

1. Uetz, P., et al.: A comprehensive analysis of protein-protein interactions in Saccharomyces cerevisiae. Nature. **403**(6770), 623–627 (2000)
2. Bondos, S.E., Tan, X.X., Matthews, K.S.: Physical and genetic interactions link hox function with diverse transcription factors and cell signaling proteins. Mol. Cell. Proteomics. **5**(5), 824–834 (2006)
3. Rain, J.C., et al.: The protein-protein interaction map of Helicobacter pylori. Nature. **409**(6817), 211–215 (2001)
4. Singh, R., Xu, J., Berger, B.: Struct2net: integrating structure into protein-protein interaction prediction. Pac. Symp. Biocomput. **11**, 403–414 (2006)
5. Sun, T., et al.: Sequence-based prediction of protein-protein interaction using a deep-learning algorithm. BMC Bioinformatics. **18**(1), 277 (2017)
6. Singh, R., Xu, J., Berger, B.: Global alignment of multiple protein interaction networks with application to functional orthology detection. Proc. Natl. Acad. Sci. USA. **105**(35), 12763–12768 (2008)
7. Shi, X., et al.: BMRF-MI: integrative identification of protein interaction network by modeling the gene dependency. BMC Genomics. **16**(1 7), S10 (2015)
8. Han, J.D., et al.: Effect of sampling on topology predictions of protein-protein interaction networks. Nat. Biotechnol. **23**(7), 839–844 (2005)
9. Hashemifar, S., Xu, J.: HubAlign: an accurate and efficient method for global alignment of protein-protein interaction networks. Bioinformatics. **30**(17), i438–i444 (2014)
10. Hashemifar, S., et al.: ModuleAlign: module-based global alignment of protein-protein interaction networks. Bioinformatics. **32**(17), i658–i664 (2016)

11. Hashemifar, S., Huang, Q., Xu, J.: Joint alignment of multiple protein-protein interaction networks via convex optimization. J. Comput. Biol. **23**(11), 903–911 (2016)
12. Dutkowski, J., et al.: A gene ontology inferred from molecular networks. Nat. Biotechnol. **31**(1), 38–45 (2013)
13. Rinner, O., et al.: An integrated mass spectrometric and computational framework for the analysis of protein interaction networks. Nat. Biotechnol. **25**(3), 345–352 (2007)
14. Zhang, A.: Protein interaction networks: computational analysis, p. 8. Cambridge University Press, Cambridge; New York (2009)
15. Kalaev, M., et al.: NetworkBLAST: comparative analysis of protein networks. Bioinformatics. **24**(4), 594–596 (2008)
16. Ciriello, G., et al.: AlignNemo: a local network alignment method to integrate homology and topology. PLoS One. **7**(6), e38107 (2012)
17. Klau, G.W.: A new graph-based method for pairwise global network alignment. BMC Bioinformatics. **10**(**1**), S59 (2009)
18. Singh, R., Xu, J., Berger, B.: Global alignment of multiple protein interaction networks. Pac. Symp. Biocomput. **13**, 303–314 (2008)
19. Kuchaiev, O., Przulj, N.: Integrative network alignment reveals large regions of global network similarity in yeast and human. Bioinformatics. **27**(10), 1390–1396 (2011)
20. Patro, R., Kingsford, C.: Global network alignment using multiscale spectral signatures. Bioinformatics. **28**(23), 3105–3114 (2012)
21. Chindelevitch, L., et al.: Optimizing a global alignment of protein interaction networks. Bioinformatics. **29**(21), 2765–2773 (2013)
22. Neyshabur, B., et al.: NETAL: a new graph-based method for global alignment of protein-protein interaction networks. Bioinformatics. **29**(13), 1654–1662 (2013)
23. Sahraeian, S.M., Yoon, B.J.: SMETANA: accurate and scalable algorithm for probabilistic alignment of large-scale biological networks. PLoS One. **8**(7), e67995 (2013)
24. Hu, J., Kehr, B., Reinert, K.: NetCoffee: a fast and accurate global alignment approach to identify functionally conserved proteins in multiple networks. Bioinformatics. **30**(4), 540–548 (2014)
25. Alkan, F., Erten, C.: BEAMS: backbone extraction and merge strategy for the global many-to-many alignment of multiple PPI networks. Bioinformatics. **30**(4), 531–539 (2014)
26. Gligorijevic, V., Malod-Dognin, N., Przulj, N.: Fuse: multiple network alignment via data fusion. Bioinformatics. **32**(8), 1195–1203 (2016)
27. Liao, C.S., et al.: IsoRankN: spectral methods for global alignment of multiple protein networks. Bioinformatics. **25**(12), i253–i258 (2009)
28. Flannick, J., et al.: Automatic parameter learning for multiple local network alignment. J. Comput. Biol. **16**(8), 1001–1022 (2009)
29. Kuchaiev, O., et al.: Topological network alignment uncovers biological function and phylogeny. J. R. Soc. Interface. **7**(50), 1341–1354 (2010)
30. Saraph, V., Milenkovic, T.: MAGNA: maximizing accuracy in global network alignment. Bioinformatics. **30**(20), 2931–2940 (2014)
31. Das, J., Yu, H.: HINT: High-quality protein interactomes and their applications in understanding human disease. BMC Syst. Biol. **6**, 92 (2012)
32. Sahraeian, S.M., Yoon, B.J.: A network synthesis model for generating protein interaction network families. PLoS One. **7**(8), e41474 (2012)
33. Kim, W.K., Marcotte, E.M.: Age-dependent evolution of the yeast protein interaction network suggests a limited role of gene duplication and divergence. PLoS Comput. Biol. **4**(11), e1000232 (2008)
34. Ohbayashi, F., et al.: Correction of chromosomal mutation and random integration in embryonic stem cells with helper-dependent adenoviral vectors. Proc. Natl. Acad. Sci. USA. **102**(38), 13628–13633 (2005)
35. Kanehisa, M., et al.: KEGG for integration and interpretation of large-scale molecular data sets. Nucleic Acids Res. **40**(Database issue), D109–D114 (2012)

36. Schlicker, A., et al.: A new measure for functional similarity of gene products based on gene ontology. BMC Bioinformatics. **7**, 302 (2006)
37. Bader, G.D., Hogue, C.W.: An automated method for finding molecular complexes in large protein interaction networks. BMC Bioinformatics. **4**, 2 (2003)
38. Yu, H., et al.: The importance of bottlenecks in protein networks: correlation with gene essentiality and expression dynamics. PLoS Comput. Biol. **3**(4), e59 (2007)
39. Heino, J., Calvetti, D., Somersalo, E.: Metabolica: a statistical research tool for analyzing metabolic networks. Comput. Methods Prog. Biomed. **97**(2), 151–167 (2010)
40. Zotenko, E., et al.: Why do hubs in the yeast protein interaction network tend to be essential: reexamining the connection between the network topology and essentiality. PLoS Comput. Biol. **4**(8), e1000140 (2008)
41. Wang, E., Lenferink, A., O'Connor-McCourt, M.: Cancer systems biology: exploring cancer-associated genes on cellular networks. Cell. Mol. Life Sci. **64**(14), 1752–1762 (2007)
42. Kerrien, S., et al.: The IntAct molecular interaction database in 2012. Nucleic Acids Res. **40**(Database issue), D841–D846 (2012)
43. Park, Y., Bader, J.S.: Resolving the structure of interactomes with hierarchical agglomerative clustering. BMC Bioinformatics. **12**(**Suppl 1**), S44 (2011)
44. Schoen, F.: Combinatorial optimization: polyhedra and efficiency (algorithms and combinatorics). J. Oper. Res. Soc. **55**(9), 1018–1019 (2004)
45. Ashburner, M., et al.: Gene ontology: tool for the unification of biology. The Gene Ontology Consortium. Nat Genet. **25**(1), 25–29 (2000)
46. Parrish, J.R., et al.: A proteome-wide protein interaction map for Campylobacter jejuni. Genome Biol. **8**(7), R130 (2007)
47. Peregrin-Alvarez, J.M., et al.: The modular organization of protein interactions in Escherichia coli. PLoS Comput. Biol. **5**(10), e1000523 (2009)
48. Boyd, S.P.: Distributed optimization and statistical learning via the alternating direction method of multipliers, p. 126. Now Publishers, Hanover, MA (2011)
49. Huang, Q.X., Guibas, L.: Consistent shape maps via semidefinite programming. Computer Graphics Forum. **32**(5), 177–186 (2013)
50. Kumar, M.P., Kolmogorov, V., Torr, P.H.S.: An analysis of convex relaxations for MAP estimation of discrete MRFs. J. Mach. Learn. Res. **10**, 71–106 (2009)
51. Huang, Q.X., Koltun, V., Guibas, L.: Joint shape segmentation with linear programming. ACM Trans. Graph. **30**(6), 125 (2011)

Chapter 4
BioFabric Visualization of Network Alignments

Rishi M. Desai, William J. R. Longabaugh, and Wayne B. Hayes

Abstract
Background: Dozens of global network alignment algorithms have been developed over the past 15 years. Effective network visualization tools are lacking and would enhance our ability to gain an intuitive understanding of the strengths and weaknesses of these algorithms.
Results: We have created a plugin to the existing network visualization tool BioFabric, called *VISNAB: Visualization of Network Alignments using BioFabric*. We leverage BioFabric's unique approach to layout (nodes are horizontal lines connected by vertical lines representing edges) to improve the understanding of network alignment performance. Our visualization tool allows the user to clearly spot deficiencies in alignments that cannot be detected by simply evaluating and comparing standard numerical topological measures such as the Edge Coverage (*EC*) or Symmetric Substructure Score (S^3). Furthermore, we provide new automatic layouts that allow researchers to identify problem areas in an alignment. Finally, our new definitions of *node groups* and *link groups* that arise from our visualization technique allow us to also introduce novel numeric measures for assessing alignment quality.
Conclusions: Our new approach to visualize network alignments will allow researchers to gain a new, and better, understanding of the strengths and shortcomings of the many available network alignment algorithms.

Keywords Network · Visualization · Alignment

Address correspondence on visualization to William J. R. Longabaugh; correspondence on alignment to Wayne B. Hayes.

R. M. Desai · W. B. Hayes (✉)
Department of Computer Science, University of California, Irvine, CA, USA
e-mail: whayes@uci.edu

W. J. R. Longabaugh
Institute for Systems Biology, Seattle, WA, USA
e-mail: wlongabaugh@systemsbiology.org

© Springer Nature Switzerland AG 2021

B.-J. Yoon, X. Qian (eds.), *Recent Advances in Biological Network Analysis*,
https://doi.org/10.1007/978-3-030-57173-3_4

4.1 Background

Network alignment is the process of finding a mapping between the nodes of two networks. Alignments have been used in social networks to identify social structure and de-anonymize supposedly anonymous networks [1], brain connectomes to aid the creation of a standardized brain atlas [2, 3], and biomolecular networks to identify functional relationships and to transfer information between species [4]. The increasing availability of protein-protein interaction (PPI) networks over the past two decades has spurred the development of dozens of new network alignment algorithms. Particularly, the alignments of PPI networks reveal key insights in protein function and similarity [4], which in turn offers better understanding of mechanisms of human diseases [5] and the process of aging in humans [6]; this may enable the transfer of biological information across species.

The BioFabric network visualization tool was created to help visualize and analyze large and complex networks [7]. While the traditional way is drawing nodes as dots and edges as lines (a method we refer to as *node–link diagrams*), BioFabric depicts nodes as *horizontal lines*, while edges are vertical lines; small squares at the top and bottom of an edge denote the nodes (horizontal lines) the edge connects. There is one node line per row, and one edge line per column, arranged on a strictly regular grid. Thus, edges *never* overlap, completely eliminating the inevitable "hairball" that results from depicting dense networks with a traditional node–link diagram. Also, since links can originate and terminate anywhere along the appropriate node line segments, there is complete freedom to decide where a link is drawn. This flexibility for edge placement can be used to create many types of network visualizations that present meaningful semantic groupings of edges. The approach of BioFabric has some similarities with "visibility representations" [8, 9], and a small example of using "nodes as lines" appeared in [10]. But BioFabric does not constrain nodes to be represented as discrete blocks and does not attempt to minimize link crossings at all. Indeed, in BioFabric it is possible to create useful network visualizations even when there are millions of intersections between node and edge lines.

There have been previous tools that have tackled the problem of visualizing network alignments [11, 12]. These tools still rely on variations of node–link diagrams and consequently suffer from the fundamental problems that plague this technique: alignments become harder to visualize as they become larger and more complex. The advantages provided by our new BioFabric plugin permit an entirely new approach to visualizing network alignments.

4.1.1 Our Contributions

BioFabric's ability to group and order *both* nodes and edges in meaningful ways provides a new way to visualize and understand network alignments. Here we will introduce a new system of node and edge groupings, with a new associated layout, that provides unique insights into alignments. We also provide a new layout technique that leverages the linear ordering generated by node misalignments that can highlight problems in the alignment. Finally, we introduce some new metrics, inspired by our visualization approach, that can be used to assess the quality of an alignment.

4.2 Methodology

4.2.1 Overview and Nomenclature

Figure 4.1 illustrates how BioFabric depicts network alignments and introduces the color-based nomenclature we use in this paper. Figure 4.1a depicts the alignment using a traditional node–link representation. Nodes from the smaller *blue* network (top left) are aligned one-to-one onto nodes from the larger *red* network (top center), creating a combined network (top right) that is the union of both under one of many possible alignments. When network elements (nodes or edges) are matched in this procedure, we refer to them as *purple*. We refer to unmatched elements from the smaller network as *blue*, and unmatched elements from the larger network are called *red*.

Of course, in this example we are making the simplifying (and very common) assumption that *all* of the nodes in the smaller network are aligned to nodes in the larger network. In fact, in the most general alignment problem, we can also end up with *blue* nodes in the final alignment. In fact, VISNAB is fully capable of dealing with blue nodes in the final alignment. However, in our discussion of the technique, and in all the case studies we are discussing in this paper, we will deal with alignments with no blue nodes in the interest of simplicity. See the Appendix for additional information on how VISNAB handles blue nodes.

Figure 4.1b shows how BioFabric represents the combined network: the nodes are depicted as horizontal lines, and the edges are depicted as vertical lines. For this example, to make the comparison more concrete, this figure departs from the actual BioFabric presentation in two ways. First, the nodes here are drawn purple and red to match their counterparts in the node–link diagram; second, the different classes of edges are distinctly grouped. In BioFabric, in order to allow the network alignment to scale, node colors are cycled in a strict pattern, and edges are organized on a uniform regular grid.

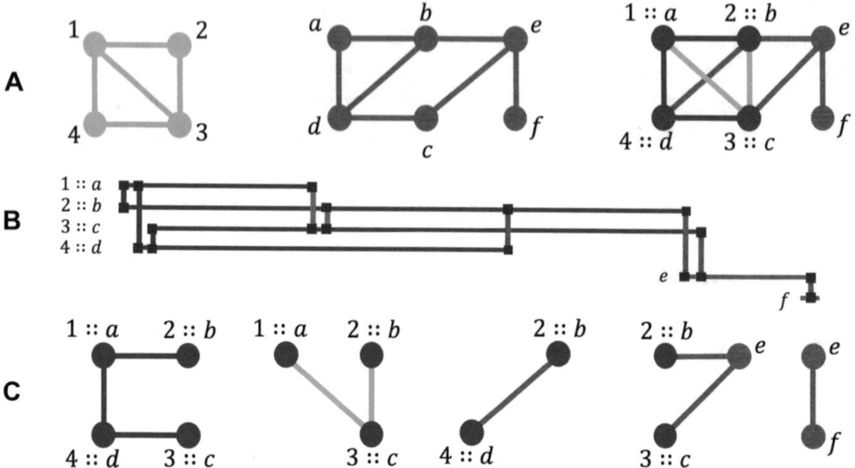

Fig. 4.1 Network alignments with BioFabric. (**a**) A traditional node–link diagram to show a simple alignment, where a smaller blue network (upper left) is aligned onto a larger red network (upper center). The resulting aligned network (upper right) represents aligned nodes and edges as purple; node 1 aligned onto a is labeled as 1::a. Elements not aligned remain blue or red. (**b**) BioFabric visualizes the same upper-right network, where nodes are drawn as *horizontal lines* and edges as *vertical lines*. Purple horizontal lines are aligned nodes; red horizontal lines are unaligned nodes. There are no blue horizontal lines because we assume every node from G_1 is aligned to some node in G_2. (**c**) Each of the five different link groups are described in the text, with the edges drawn directly under their corresponding BioFabric depiction. Not shown in this diagram are the node groups, which are not ordered in this depiction, for each node: 1::a: **(P:pBp)**; 2::b: **(P:P/pBp/pRp/pRr)**; 3::c: **(P:P/pBp/pRr)**; 4::d: **(P:P/pRp)**; e: **(R:pRr/rRr)**, f: **(R:rRr)**

4.2.2 Node and Link Groupings

Consider the *rows* in Fig. 4.1b: aligned purple nodes (1::a through 4::d) are grouped together exclusively in the top rows, while the unaligned red nodes (e and f) are grouped together exclusively in the bottom rows. Similarly, edges can be grouped in columns. As Fig. 4.1c shows, there are *five* distinct classes of edges in the aligned network. Moving from left to right, we have purple *matched* edges, blue *orphan* edges, and three distinct classes of red *untouched* edges defined based upon the color of the two nodes incident on the edge: both purple, both red, or one of each. We refer to these five distinct classes as the *link groups* for a network alignment; they are enumerated in Table 4.1, along with the symbol we use to describe each one (e.g., **pRr**). For an alignment without blue nodes, these five link groups are a partition P of the set of edges in the union of the blue and red networks (where empty sets are allowed in P). The full enumeration of the seven link groups for the blue node case is provided in the Appendix.

BioFabric's link group separation allows visual quantification of the "common topology" discovered by an alignment. For example, one could argue that a better alignment would result by rotating the blue square 90° in either direction, combining

Table 4.1 The enumeration of VISNAB's link groups for the case of an alignment with no blue nodes. The link group numbering corresponds to the numbering used for the full blue node case, given in the Appendix

Link group	Edge color	Endpoint 1	Endpoint 2	Symbol
1	Purple	Purple	Purple	P
2	Blue	Purple	Purple	pBp
5	Red	Purple	Purple	pRp
6	Red	Purple	Red	pRr
7	Red	Red	Red	rRr

Table 4.2 The enumeration of VISNAB's node groups for the case of an alignment with no blue nodes. The node group numbering corresponds to the numbering used for the full blue node case, given in the Appendix

Node group	Node color	Incident edges	Symbol
1	Purple	None	(P:0)
2	Purple	Purple	(P:P)
3	Purple	pBp	(P:pBp)
6	Purple	pRp	(P:pRp)
7	Purple	Purple, pBp	(P:P/pBp)
10	Purple	Purple, pRp	(P:P/pRp)
11	Purple	pBp, pRp	(P:pBp/pRp)
14	Purple	Purple, pBp, pRp	(P:P/pBp/pRp)
17	Purple	pRr	(P:pRr)
18	Purple	Purple, pRr	(P:P/pRr)
19	Purple	pBp, pRr	(P:pBp/pRr)
22	Purple	pRp, pRr	(P:pRp/pRr)
23	Purple	Purple, pBp, pRr	(P:P/pBp/pRr)
26	Purple	Purple, pRp, pRr	(P:P/pRp/pRr)
27	Purple	pBp, pRp, pRr	(P:pBp/pRp/pRr)
30	Purple	Purple, pBp, pRp, pRr	(P:P/pBp/pRp/pRr)
37	Red	pRr	(R:pRr)
38	Red	rRr	(R:rRr)
39	Red	pRr, rRr	(R:pRr/rRr)
40	Red	None	(R:0)

the red and blue edges in the square's interior to a single purple edge; the BioFabric layout of the resulting alignment would immediately and visually depict the elimination of the red and blue interior edges.

The two major classes of purple and red nodes can be further subdivided into *node groups* based upon the types of edges incident upon the node; these twenty groups and their symbols are enumerated in Table 4.2 for the case with no blue nodes. For example, red nodes that are not singletons (groups 37–39) can be classified as having node neighbors that are: (1) only purple (all incident edges are **pRr**), (2) purple and red (incident edges are **pRr** and **rRr**), or (3) only red (all incident edges are **rRr**). As with the link groups, for an alignment without blue nodes, these 20 groups are a partition P of the set of nodes in the union of the blue and red networks (where empty sets are allowed in P). The full enumeration of the 40 node groups for the blue node case is also provided in the Appendix.

4.2.3 Network Merge

Most attempts to visualize an alignment provide side-by-side node–link diagrams of the input networks, with edges drawn between them to depict the alignment. The result rarely provides intuition about common topology discovered by the alignment. Instead, we *merge* the input networks (cf. Fig. 4.1a, far right), characterizing the result with our node and link groups. Let $G_1 = (V_1, E_1)$ and $G_2 = (V_2, E_2)$ be two networks with $|V_1| \leq |V_2|$. We define a pairwise global alignment from G_1 to G_2 as an injective function $a : V_a \rightarrow V_2$; every node in $V_a \subseteq V_1$ is mapped to a distinct node in V_2, and V_a is allowed to be a strict subset of V_1, indicating some nodes in V_1 (the blue nodes) may not be mapped. The merged network $G_{12} = (V_{12}, E_{12})$ consists of the nodes and edges in the two networks. All aligned nodes $u \in V_a$, $v \in V_2$ with $a(u) = v$ are combined into one node n and added to V_{12}. A combined node n is labeled in the format u::v. The unaligned nodes in V_1 and V_2 are also added to V_{12}. E_{12} consists of all the edges in E_1 and E_2. An *aligned* edge is an edge $(u_1, u_2) \in E_1 : (a(u_1), a(u_2)) \in E_2$. Aligned edges (link group **P**) are only represented *once* in E_{12}. Hence, the total number of edges $|E_{12}|$ varies with different alignments.

4.2.4 New Metrics

Most topology-only network alignment approaches currently in use perform rather poorly across a wide range of biological test sets, particularly when the alignment is evaluated using **Node Correctness** (*NC*). Traditionally *NC* is the fraction of nodes $u \in V_1$ aligned correctly when the correct alignment is known; we extended *NC* for alignments with unaligned nodes $u \in V_1$. Let $a'(u)$ be $a(u)$ if $a(u)$ is defined and u otherwise. Given an alignment $a : V_a \rightarrow V_2$ and the known correct alignment $a_c : V_c \rightarrow V_2$ with $V_a, V_c \subseteq V_1$, our *NC* measure is defined as the fraction of nodes $u \in V_1 : a'(u) = a'_c(u)$.

In fact, slight differences in *NC* values, even if statistically significant, are arguably not very informative to assess pure topologically driven alignments. We demonstrate here that other ways of assessing alignment performance can provide richer insights into the behavior of different techniques.

For example, if two nodes u and v in any graph G have an identical neighbor set (excluding the edge (u, v) if it exists), then it is impossible to distinguish between them using topology alone; thus, an alignment should not be penalized for swapping

u and v if one of the two potential swaps is the "correct" one. Similarly, if k nodes all share the same neighbor set (excluding all edges between the k nodes themselves), any permutation of these k nodes should be scored equally if one of the permutations yields the "correct" alignment. This observation leads us to adapt a new **Jaccard Similarity** (*JS*) measure in the context of network alignment, as follows:

For some network $G = (V, E)$, define the neighborhood of node z_1 in G to be $N_G(z_1) = \{z_2 \in V : (z_1, z_2) \in E\}$. For nodes $x, y \in V$, let $N_G(x, y)$ be the neighborhood of x disregarding y, and let i_{xy} be a corrective term accounting for a possible edge between the two, as follows: if $y \in N_G(x)$, then $N_G(x, y) = N_G(x) - y$ and $i_{xy} = 1$, else $N_G(x, y) = N_G(x)$ and $i_{xy} = 0$. Our extended *JS* definition $\sigma_G : V \times V \rightarrow [0, 1]$ between two nodes is defined as

$$\sigma_G(x, y) = \frac{|N_G(x, y) \cap N_G(y, x)| + i_{xy}}{|N_G(x, y) \cup N_G(y, x)| + i_{xy}}.$$

Intuitively, two aligned nodes score a 1.0 if they share an identical set of neighbors (ignoring a possible self-link between them) and are therefore impossible to distinguish using topology alone, and so a topology-driven alignment score should not penalize the misalignment of such pairs. (Note that when x and y are both singletons, we define $\sigma_G(x, y) \equiv 1.0$ to avoid dividing by zero.)

Given node sets $V_a, V_c \subseteq V_1$, an alignment $a : V_a \rightarrow V_2$, and the correct alignment $a_c : V_c \rightarrow V_2$, our *JS* measure for the given alignment a, with respect to the correct alignment a_c, in the case where $|V_a| = |V_c| = |V_1|$ is defined as

$$JS(a) = \frac{1}{|V_1|} \sum_{u \in V_1} \sigma_{G_2}(a(u), a_c(u)).$$

The *JS* measure for the "blue node" case, where $|V_a|, |V_c| \leq |V_1|$, is detailed in the Appendix. Significantly, Jaccard Similarity does *not* penalize a misalignment if the two nodes are topologically indistinguishable. Note also how *JS* gives "partial credit" for alignments where the nodes have nearly the same sets of neighbors.

The importance of node and link groups to VISNAB visualizations inspired us to design two new metrics that can be used to assess the performance of an alignment in the presence of a known correct alignment ($NC = 1$) between the two networks. For this, we use the node group and link group equivalence relations over the groupings of nodes and edges in an aligned network that we have developed. We can do this by comparing how the distribution of nodes and edges in the given alignment across the groupings compares to the distribution found in the correct alignment. To this end, we have created the **Node Group Similarity** (*NGS*) and **Link Group Similarity** (*LGS*) metrics. Both *NGS* and *LGS* are calculated using a vector representing the proportion of nodes and links distributed among the twenty node groups and five link groups, respectively. For example, the *NGS* vector $\mathbf{r}$ for k node groups $\mathrm{ng}_1, \mathrm{ng}_2, \ldots, \mathrm{ng}_k$ where $|V_{12}| = |V_1| + |V_2| - |V_a|$ is

$$\mathbf{r} = \left\langle \frac{|\text{ng}_1|}{|V_{12}|}, \frac{|\text{ng}_2|}{|V_{12}|}, \ldots, \frac{|\text{ng}_k|}{|V_{12}|} \right\rangle.$$

The angular similarity between the vector of the given alignment and the vector of the known correct alignment is calculated. For vectors with only positive elements, angular similarity = $1 - 2\theta/\pi$, where θ is the angle between the vector of the alignment and the vector of the correct alignment. Angular similarity was chosen (instead of, e.g., cosine similarity) because it exhibits more rapid decay from the optimum value in the case of small angles.

All of these metrics, as well as other common measures, are displayed by VISNAB in a dialog when the user provides the correct alignment file as an optional part of the set-up steps.

4.2.5 Node and Link Group Layout Algorithm

Once we have created a merged network, we must provide an informative layout of the network that organizes the nodes and edges into a grouping framework that aids understanding the alignment. All BioFabric network layouts are fully defined by creating a linear ordering of the nodes, and a linear ordering of the edges. The default technique for laying out nodes is simple: within each node group it creates a linear ordering of the nodes using a breadth-first search from the node of highest degree, where neighboring nodes are visited in the order of highest to lowest degree. This basic technique, which uses a single queue for the breadth-first search, can be modified to create a layout that organizes the nodes using the alignment-based groupings we have defined. Specifically, we adapted it to use a *separate* queue for each of the twenty node groups. The resulting layout retains much of the same overall structure as the default algorithm, while ensuring that nodes are well-organized in each node group band. As nodes are visited during the search, they are added to the queue corresponding to their node group, and these node group queues are processed in the order listed in Table 4.2. A more complete description of the algorithm is provided in the Appendix.

Finally, if the correct alignment is provided, the user can choose to have correctly and incorrectly aligned nodes laid out separately in different node groups. The user can choose the criterion for the correct alignment to be based either on the traditional *NC* measure or on our *JS* measure. If *JS* is chosen, the user can set the threshold value $\beta \in [0, 1]$, so, in the case where $|V_a| = |V_c| = |V_1|$, an aligned node u::v is denoted correct if $\sigma_{G_2}(v, a_c(u)) \geq \beta$. If *JS* is chosen, the condition for correctly aligned nodes for the "blue node" case, where $|V_a|, |V_c| \leq |V_1|$, is detailed in the Appendix.

The default edge layout algorithm, using a slightly modified version of the existing link group feature described in [7], can be used to organize the edges into the five link groups. The new modification allows edges that are tagged with a particular relation to be grouped contiguously on a *per-network* basis; previously, this could only be done on a *per-node* basis. Using this new modification, one of five tags signifying the edge's link group (Table 4.1) is assigned to each edge in the network. The edges are then partitioned based on these tags into five link groups, like those in Fig. 4.1. Consequently, the upper-left corner of the layout contains the purple nodes that are either singletons or have only purple edges (**P:P**).

4.2.6 *Alignment Cycle Layout for Self-alignments*

A common approach for assessing the performance of alignment algorithms is to align a subset of a network to itself, where the larger network contains the same number of nodes but a larger number of edges [4, 13]. To allow researchers to better understand these analyses, we developed a special network alignment layout method that highlights alignment problems. Our technique leverages the fact that if a network is aligned to itself, then it can be viewed as a set of *cycles* each covering the set of *nodes*. Correctly aligned nodes will have a cycle length of one, $u \rightarrow u$, but a misalignment of $u \rightarrow v$, and $v \rightarrow u$, will create a cycle of length two. More severe misalignments create longer cycles. Figure 4.2 shows a small example of this approach.

Since a node layout in BioFabric is just a linear ordering of nodes, these alignment cycles provide a natural basis for specifying this ordering; the resulting layout vividly portrays the nature of the alignment. We start by placing a node in the first row and then use the alignment cycle containing that node to specify the row ordering of all remaining nodes in the cycle. The decision of how to order these cycles is again derived from a modification of BioFabric's breadth-first search default layout algorithm [7]. For the alignment cycle layout, *all* the nodes in the cycle for a neighbor are simply placed first before the next neighbor is processed.

Although this technique is easiest to picture when a network is aligned to itself, the method has been generalized to handle network alignments from one network to another with *more* nodes. In this case, the alignment creates a set of *paths* instead of cycles, since not all nodes in the large network will map to a node in the small one. With paths, the algorithm places all nodes belonging to a path when any node in the path is first encountered in the search. Finally, the user can provide a mapping if the nodes in the two networks are in different namespaces (e.g., one network uses Saccharomyces Genome Database (SGD) IDs for node names, while the other network uses ENTREZ IDs).

4.2.7 *Interface with BioFabric*

BioFabric Version 2 (now in beta release) provides a new plug-in architecture that allows developers to add new features to the program. A plug-in is written in Java, and when the .jar file containing the compiled code is placed in a directory specified

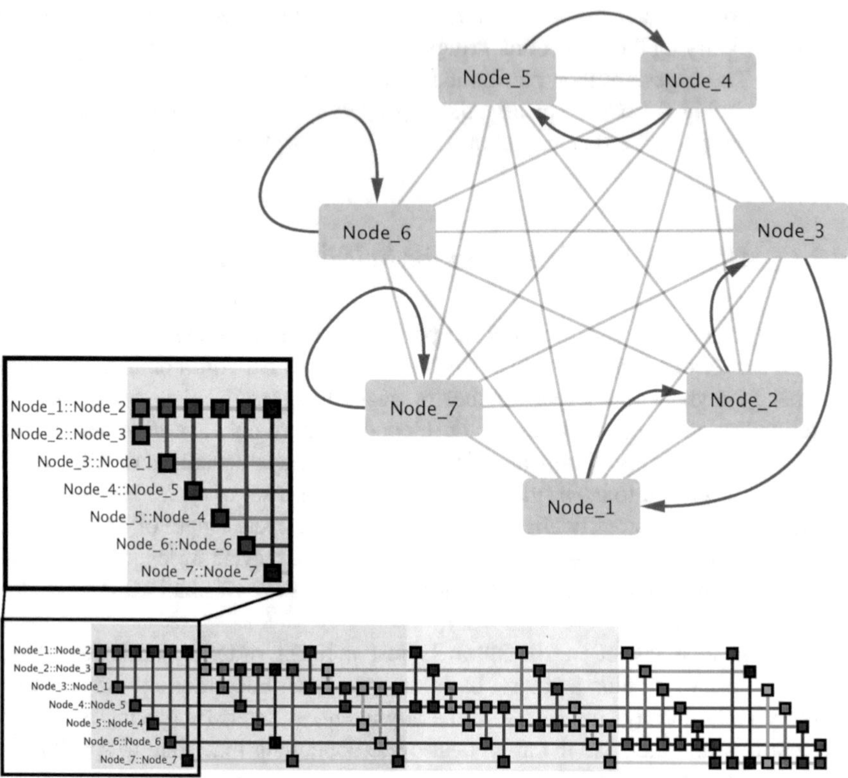

Fig. 4.2 Understanding the Alignment Cycle layout. A network consisting of seven nodes has been aligned to itself (top); the red directed edges indicate the results of the alignment, e.g., *Node_1* was incorrectly aligned to *Node_2*, while *Node_7* was correctly aligned. As shown, this results in a set of cycles (four in this example), which is then used to create the linear ordering needed for the BioFabric layout (bottom) that places aligned nodes in adjacent rows; detail on lower left shows node labels. The Alignment Cycle layout highlights incorrect alignment cycles with alternating colored blocks, as shown (block and node zone labels omitted)

by the user, the new functionality is available in BioFabric's **Tools** menu. Our VISNAB plug-in uses this new architecture to provide the new functions that allow the user to load in two networks, an alignment file, and an optional correct reference alignment, and then process and lay out the desired view of the alignment. The plug-in also uses the new node and link annotation feature that is now being introduced in BioFabric Version 2. This feature allows the user to specify spans of nodes or links that are then highlighted by drawing colored rectangles in the background.

4.3 Results

4.3.1 Case Study I: Simple Network Comparison

VISNAB can be used to directly compare two networks when aligning a network to itself. An example of this is where the nodes are proteins from the same organism and the networks are from two different studies. Our new node and link group definitions can provide visual insights into how the two networks compare. Notably, this technique can also be used to compare *any* two networks, as long as a 1:1 mapping of nodes between the two networks is provided.

Figure 4.3 shows a "correct" alignment between two datasets describing the *S. cerevisiae* (baker's yeast) PPI network. The smaller *Yeast 2K* is derived from a network generated from data in [13] and used in [4] and contains 2390 nodes and 16,127 edges, while the larger *SC* (BioGRID (v3.2.101, June 2013) [14]) contains 5831 nodes and 77,149 edges. The networks are "aligned" based upon simply matching node names. Note these networks used different protein naming conventions; the Appendix provides details on how we created this correct alignment.

The top of Fig. 4.3 shows the entire merged network, while the detail below it displays the subnetwork with purple and blue edges. The lowest detail shows the

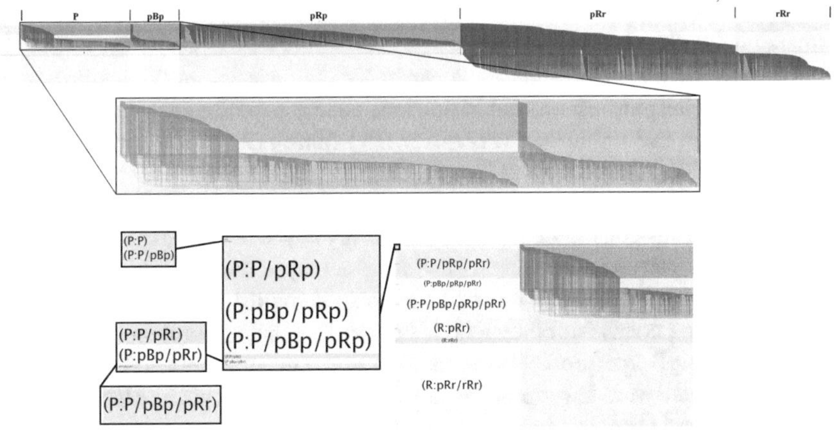

Fig. 4.3 A direct comparison of network *Yeast 2K* [13] to *SC* [14]. Top: Full network view. Middle: Detail of the subnetwork with purple (**P**) and blue (**pBp**) edges. Bottom: Detail of the node group assignments. The five link groups are denoted by alternating bands of light and dark shading; the node groups are each assigned fixed colors from BioFabric's standard palette. Outsize tags for each link group have been added here for clarity along the top of the network; outsize tags for the largest node groups have been added here for clarity as well. As described in the text, common topological measures for alignment quality such as S^3, EC, and ICS can be visually estimated using the *widths* of each link group. For example, S^3 is the width of the leftmost link group **P** over the sum of the leftmost three link groups **P**, **pBp**, and **pRp**, i.e., about 1/4 (actual value: 0.248). Node groups can be used to gain insights into how the networks compare; see text for details

distribution of nodes into our node groups, which are denoted by the horizontal colored bands. This display uses a recently added BioFabric Version 2 feature: node and link annotation displays.

Note how the five link groups and twenty node groups quickly reveal how these two networks compare. We immediately see that the number of nodes in the larger network is more than double the number in the original, as the purple nodes are in the colored bands down through the **(P:P/pBp/pRp/pRr)** (peach colored) node group; the node below that are red nodes only in the larger network. The edges in the two leftmost shaded link group bands (**P** and **pBp**) represent *exactly* the links in the *Yeast 2K* network. If the larger network were to be treated as a later, more complete, and more accurate survey, then the edges in group **pBp** represent the *false positives* from the earlier survey. The red edges in middle link group **pRp** are new edges from the more recent survey between old nodes from the original survey, while the edges in the right 40%, in the **pRr** and **rRr** link groups, represent new edges incident on at least one new red node.

Studying the relative sizes of the node groups also provides insights into how the networks differ. One thing to note right away is that the six node groups we do not expect to appear are in fact absent: **(P:0)**, **(P:pBp)**, **(P:pRp)**, **(P:pRr)**, **(P:pRp/pRr)**, and **(R:0)**. We expect this because these protein–protein interaction networks do not have any singleton nodes. Thus, for example, a node in the **(P:pBp)** class would require that the node be a singleton in the larger network. Conversely, a purple node with only red edges would require that the node be a singleton in the smaller network.

Another significant fact that is apparent by the node group distribution is that **over 90%** of the purple nodes are in the three node groups that have red edges going to both purple and red nodes: **(P:P/pRp/pRr)** (pink), **(P:pBp/pRp/pRr)** (powder blue), and **(P:P/pBp/pRp/pRr)** (peach). In contrast, there are only a tiny number of aligned nodes that have red edges going to *only* aligned or *only* unaligned nodes. Again, this seems consistent with a wider survey that would be expected to mostly find new interactions between both previously surveyed proteins and between those proteins and new proteins. The red nodes also show the same pattern of mostly (~80%) having nodes with interactions to both new and old proteins. Finally, note how group **(P:pBp/pRp/pRr)** is the smallest of the major purple node groups. This tells us that there were fewer proteins in the original network where all the old edges were considered false positives, again a result that makes sense.

4.3.2 Case Study II: Visualizing Common Topological Measures

There are several measures to assess the topological quality of an alignment. Depending on the context, these measures quantify how much topological similarity between the two PPI networks an alignment has exposed. Let $E_a = \{(u_1, u_2) \in E_1 : (a(u_1), a(u_2)) \in E_2\}$ denote the edges in G_1 aligned to edges in G_2, i.e., purple edges. The ***EC*** measure (variously called Edge Coverage, Conservation, Correctness, or Correspondence by different authors) is the fraction of edges in the smaller network that are aligned to edges in the larger network: $EC(a) = |E_a|/|E_1|$. Let $\hat{E}_a = \{(v_1, v_2) \in E_2 : \exists u_1, u_2 \in V_1 \wedge a(u_1) = v_1 \wedge a(u_2) = v_2\}$ denote the set of edges of G_2 induced on its aligned nodes. **Induced Conserved Structure** (*ICS*) is the ratio of aligned edges to induced edges: $ICS(a) = |E_a|/|\hat{E}_a|$. However, *EC* and *ICS* have the shortcoming that they can be high if the alignment maps sparse regions of one network to dense regions of the other. In the extreme, if G_2 is a clique, any alignment has $EC = 1$. To overcome this drawback, [15] devised the **Symmetric Substructure Score** (S^3), the ratio of aligned edges to all edges with aligned endpoint nodes:

$$S^3(a) = \frac{|E_a|}{|E_1| + |\hat{E}_a| - |E_a|} = \frac{\text{purple edges}}{\text{edges of all colors between purple nodes}}.$$

Since BioFabric lays out node rows and link columns on an absolute regular grid, the *proportions* of the widths of various node and link groups allow the user to rapidly visualize these topological metrics at a glance. For example, S^3 can be estimated by comparing the *width* of the first link group (purple edges) over the *width* of the first three link groups (all links between purple nodes):

$$S^3 = \frac{|\mathbf{P}|}{|\mathbf{P}| + |\mathbf{pBp}| + |\mathbf{pRp}|}.$$

Looking at the top view of Fig. 4.3, we visually estimate that this alignment has an S^3 score of about 1/4 (actual value: 0.248). In the same fashion, *EC* is the width of the purple link group over the width of the purple and blue link groups:

$$EC = \frac{|\mathbf{P}|}{|\mathbf{P}| + |\mathbf{pBp}|}.$$

We can visually estimate *EC* to be about 2/3 (actual value: 0.689). Finally, *ICS* is the width of the purple link group over width of the purple group and red group with purple endpoint nodes:

$$ICS = \frac{|\mathbf{P}|}{|\mathbf{P}| + |\mathbf{pRp}|}.$$

We can visually estimate *ICS* to be around 3/10 (actual value: 0.280). Thus, BioFabric's node and link groups provide intuition on abstract topological measures.

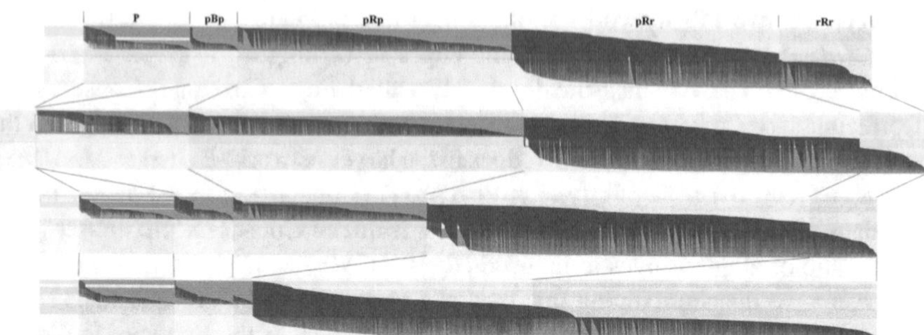

Fig. 4.4 Visually assessing the performance of different objective functions with four alignments between *Yeast 2K* and *SC*. First row: gold-standard; second row: $1.0 * I$ (Importance-only); third row: mixed with $(0.03 * S^3) + (0.97 * I)$; fourth row: $1.0 * S^3$. We have drawn lines that follow each link group from the first row's alignment to the respective link group in each successive row's alignment. The alignment in the second row is noticeably wider than the others because aligned **P** edges represent *two* edges from the original network pair, and this alignment has very few **P** edges

4.3.3 Case Study III: Visualizing Performance of Objective Functions

Choosing the best objective function to apply to a particular network alignment is currently a black art, and providing the researcher with a visual tool to assist in understanding the properties and performance of various objective functions helps to gain insights into this difficult problem.

Indeed, BioFabric's presentation of network alignment allows the researcher to do just this. Figure 4.4 depicts four alignments between the *Yeast 2K* and *SC* networks previously introduced. The first row of the figure is the same gold-standard, correct alignment shown in Fig. 4.3. For the other alignments, we used SANA [16], which allows us to select the objective functions we wish to use to create an alignment. These other three alignments were generated by running SANA for 10 h each (long enough for the random search algorithm to converge to a near-optimal score of the chosen objective function), optimizing the following objective functions: the second-row alignment was generated with an objective function utilizing only **Importance** (I) [17]; the third with 0.03 weight given to S^3 and 0.97 weight given to I; and the fourth utilizing only S^3.

In addition to viewing the BioFabric plots of the four alignments, we can also look at our new *NGS*, *LGS*, and *JS* metrics for them, as well as the traditional *NC*, S^3, and Resnik semantic similarity scores [18–20] calculated by FastSemSim [21]. These scores are shown in Table 4.3. The highest S^3 value and the highest *NC* value are often the measures used to identify the "best" alignment. As the table shows, the pure S^3 version provides the highest of both of these scores. Yet this "best" *NC* score is only 2.1%, which is not a very inspiring result. Even our new *JS* score of 6.9%, which is tolerant of mismatches that are topologically similar, is still very low.

Table 4.3 Scores for the four alignments between *Yeast 2K* and *SC*

Alignment	NGS	LGS	NC	JS	S^3	Resnik
Correct	1.00	1.00	1.00	1.00	0.25	9.63
$1.0 * I$	0.61	0.79	0.00042	0.021	0.0043	3.16
$0.03 * S^3 + 0.97 * I$	0.87	0.80	0.018	0.057	0.27	3.62
$1.0 * S^3$	0.64	0.46	0.021	0.069	0.55	3.61

Simply *looking* at the four BioFabric plots in Fig. 4.4 makes us question calling the pure S^3 version the "best" alignment of these two networks. Recalling that S^3 is the width of the first link group over the width of the first three link groups, we can instantly see that (1) the pure S^3 version has a much higher value than any other alignment, even the correct alignment, and (2) it achieves this by creating an alignment that forces *far* more edges into the two rightmost link groups **pRr** and **rRr**. Particularly worrisome is the very large increase in the size of **rRr** compared to the correct alignment; this set represents the edges and nodes that have been completely omitted from the alignment.

This fact that high-degree nodes were relegated to the red untouched nodes provides a clue to a way to improve the situation. Perhaps using Importance (in which the highest degree nodes tend to align to each other) as an objective function can remedy the problem. This approach gives rise to the "Pure Importance" version in row two of Fig. 4.4. Indeed, this version has link groups **pRr** and **rRr** on the right end that appear to be *much* closer in size to the correct alignment, and the center link group **pRp** is closer to the correct alignment as well, albeit larger. But the leftmost group of purple **P** edges is so thin that it is almost nonexistent. Hence, pure Importance yields an S^3 value of merely 0.0043.

So, perhaps a simple linear mixture of the two objective functions can provide a compromise between these two extremes? To investigate this, we ran a series of alignment runs that used a linear mixture of the two objective functions, and nine combinations were scored. The scores for all nine are included in the Appendix. The mixture $(0.03 * S^3) + (0.97 * I)$ is shown in the third row and provides a reasonable *visual match* of the gold-standard alignment in terms of the distribution of edges between the five link groups. This visual similarity is also borne out by the metrics shown in Table 4.3.

Scanning across a variety of mixture values, this particular mix produced an alignment with a high *NGS*, a reasonably high *LGS*, a nonzero *NC*, and a respectable *JS* value. It also has the closest S^3 value to the gold standard, plus the highest functional similarity, i.e., the most biological relevance per the Resnik score, of all the alignments, though only marginally.

Another way to compare performance is by studying the node group distribution under the different objective functions, as shown in Fig. 4.5. Notably, the "best" mixed alignment's four largest node groups are the same as the four largest node groups of the correct alignment. The two largest purple node groups of each, **(P:P/pRp/pRr)** (pink) and **(P:P/pBp/pRp/pRr)** (peach), show a rich mixture of

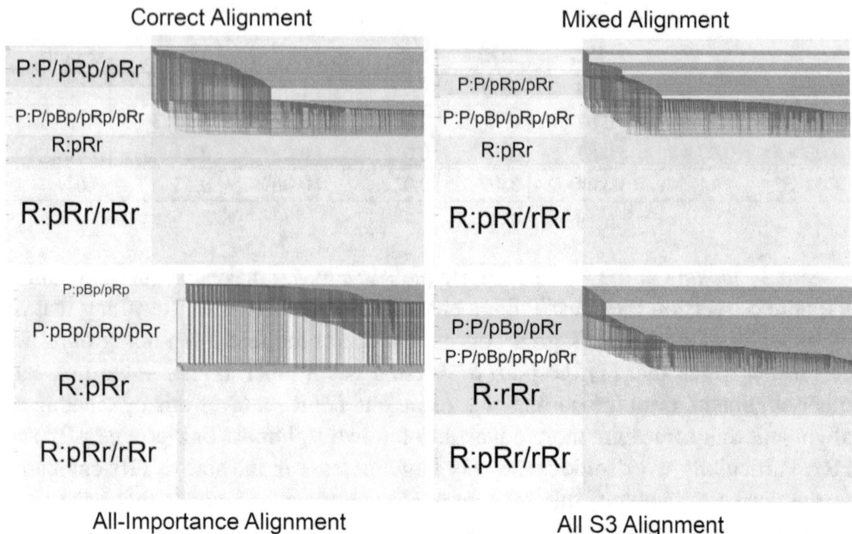

Fig. 4.5 Visualization of node groups can also provide insights into understanding the performance of objective functions. Here we show the far left side of the four alignments in Fig. 4.4, with the four largest node groups for each shown with added outsize tags for clarity. Note that the four largest node groups of the "best" mixed alignment (upper right) are the same as those of the correct alignment (upper left), though there are size differences

incident purple, red, and (in the latter group) blue edges. A significant difference between the two is the thicker bands above the pink **(P:P/pRp/pRr)** region for the mixed alignment compared to the correct alignment. The majority (76%) of nodes in those bands have no **pRr** incident edges, which reflect the fact that although there are *more* **pRr** edges in the mixed alignment (see Fig. 4.4), those edges are concentrated across a *smaller* fraction of the purple nodes. Specifically, 23% of purple nodes in the mixed alignment have no **pRr** incident edges, compared to just 5% in the correct case. Table 4.10 in the Appendix provides the exact sizes of the different node groups for the four alignments.

4.3.4 Case Study IV: Finding Protein Cluster Misalignments

While looking at metric values for a particular alignment can provide some general idea of how an alignment performs, there is value in being able to quickly spot major problems in an alignment where a gold-standard alignment is available and then be able to understand what the problem is.

Figure 4.6 shows a yeast network, *Yeast0*, of 1004 nodes and 8323 edges aligned to a network with the same set of nodes, but with 20% more (9987) edges, *Yeast20*. This network is part of the *noisy yeast* variations on the network in [13]. The edges

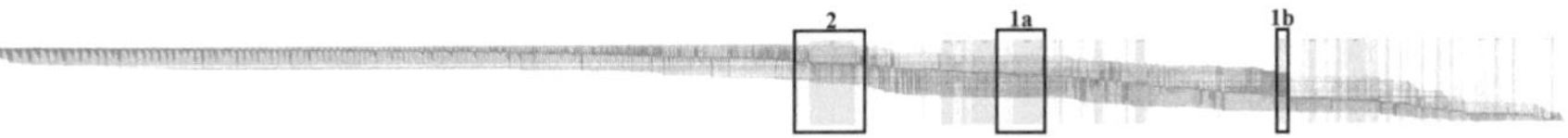

Fig. 4.6 Full view of the problematic *Yeast0* to *Yeast20* alignment, arranged using the Alignment Cycle layout. BioFabric *shadow links* [7] are turned on. Alignment cycles of incorrectly aligned nodes are indicated with alternating orange and green blocks. The numbered boxes highlight the (noncontiguous) sections containing the one- and two-cycle cluster artifacts shown in Fig. 4.7 and discussed in the text

in the smaller network are a strict subset of the larger set. This dataset has been used in several previous studies [15, 16]. The particular alignment we are using here was part of an ensemble generated by SANA [16] but was unusually poor compared to other alignments in the ensemble; this leads us to ask what was so different about this poor alignment. This visualization is created using the same network merge technique described in **Network Merge**, above. The nodes in this network are labeled to represent the alignment result, e.g., *a*::*a* for a correct alignment, and *a*::*b* for an incorrect alignment. Edges are labeled **P** (for purple edges), **pBp** (blue), or **pRp** (red).

To be able to spot alignment problems, we view the network using our new Alignment Cycle layout method, described above. This layout does not partition nodes into node groups but still generates link groups. Since the set of nodes is the same for both networks, there are no red nodes, so only three of the five link groups are present. For this type of presentation, we employ BioFabric's *per-node* link group layout to separate links into the three distinct groups within each node's dedicated *node zone*. This is in contrast to the *per-network* approach we used previously, which creates single global regions for each link group.

The large uncolored stretches of Fig. 4.6 indicate that most of the nodes—particularly high-degree ones—are correctly aligned. However, cycles of misalignment, even those of length two, are shown with alternating orange and green blocks, generated using the link annotation feature. In order to understand these misalignments better, we zoom into each region to study them more closely.

One interesting region includes a misalignment cycle containing 15 nodes, as shown in Fig. 4.7A. A node–link diagram version created using Cytoscape [22] is shown, where the protein–protein interactions are blue, and directed red edges represent the alignments. The BioFabric representation is shown at the top, with an inset detail shown as well. In addition to the large 15-node misalignment cycle on the left half of the node–link diagram, there are also correct alignments and small cycles of length two and three.

This is clearly a single protein cluster; in fact, it is some of the protein components of the mitochondrial ribosomal protein of the large subunit. (Note that all characterizations of proteins in this paper were obtained from the SGD [23].) While not fully connected, it does approximate a clique, so the misalignments found here are unsurprising, given the objective function uses topology alone.

Crucially, we can instantly interpret the context of the misalignment with the BioFabric version of the network: the repeating pattern of "edge wedges" is the

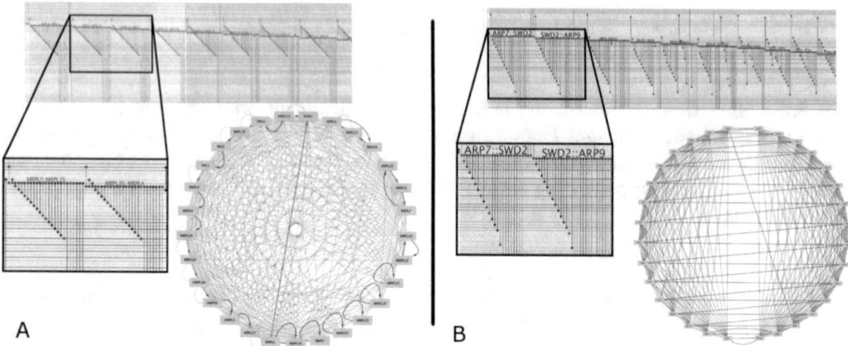

Fig. 4.7 BioFabric allows the user to spot major alignment problems; in this case, we can see how two entire protein complexes were misaligned. In section **A**, left, we show proteins of the mitochondrial ribosomal large subunit; the traditional node–link diagram for this complex is shown with protein–protein interaction edges (colored blue) and alignments shown as directed edges (colored red). The misalignments (i.e., alignment edges that are not self-loops) are to be expected, given the high topological similarity of the nodes, but nodes in this cluster are aligned with other nodes in the same cluster. Note in the BioFabric view (top left, with detail in lower left), the repeating "edge wedge" pattern with the 45° angle is the canonical representation for a clique. The ordering of the nodes using alignment cycles has no visible effect when node misalignments are intracluster. In section **B**, compare the shape of the "edge wedges" in this BioFabric view to the one in **A**; note the wedges have a 60° (not 45°) angle and the edges for each node are incident on *alternating* and *disjoint* sets of node rows in a context where misaligned nodes are in adjacent rows. This is an unmistakable visual cue that there has been a misalignment between two completely distinct protein complexes. Note how in the node–link diagram (lower right) all the red node alignment links cross back and forth between the two sides. The left side has proteins of the CPF cleavage and polyadenylation factor, while the right side has components of the RSC chromatin remodeling complex

telltale sign in BioFabric of a clique when the nodes in the clique are laid out contiguously. Remembering that the order of the node rows has been determined using the order of the alignment cycles, we see that in this case the node order still maintains the typical clique pattern; thus, the misalignments are actually *contained within the clique*. Since nodes within a clique are topologically indistinguishable, their misalignment is unavoidable (unless more information such as gene sequence data is used in the alignment objective function) and should not be penalized for an alignment, which is driven by topology alone.

A more profound alignment problem is shown in Fig. 4.7B. This misalignment involves 32 nodes in three alignment cycles. As the red edges in the node–link diagram show, the alignment has incorrectly aligned one *entire* protein complex with a completely different complex! On the left are proteins of the CPF cleavage and polyadenylation factor, while on the right are the components of the RSC chromatin remodeling complex.

This problem stands out in the BioFabric version, where the user typically studies and compares the "edge wedge" shapes for nodes to better understand the network structure. While a typical clique pattern (as shown in Fig. 4.7A) has edge wedges with a 45° lower margin, these wedges have a steeper, 60° lower margin; the edges for each node are actually incident on *alternating* nodes in this run. The explanation of this pattern is that the nodes belonging to the two separate protein clusters are intercalated in this run of nodes that have been laid out using the alignment cycles. Thus, our Alignment Cycle layout technique makes this problem stand out just by looking at it.

Even if one would argue that the traditional node–link diagram shows these effects well, the views shown had to be meticulously handcrafted to illustrate the structure once we knew what we were looking at, while the BioFabric representations were automatically laid out. As we show in the Appendix, even a complex misalignment involving a cycle of *four* protein clusters can be quickly spotted.

4.4 Discussion

Given the difficulty of viewing the quality and characteristics of network alignments using traditional node–link diagrams, it is unsurprising that researchers have concentrated on evaluating alignments by comparing and optimizing certain numerical metrics. But to be able to gain some broad intuition about network alignments, having effective and well-organized visualization techniques can provide huge benefits in understanding the problems.

While still providing the deep intuition of the node–link approach (albeit with the nodes depicted as lines), BioFabric additionally gives the user the ability to group and order *both* nodes and edges in meaningful ways. This gives researchers a new way to tackle the problem of understanding network alignments. The organized grouping of *edges* that is made possible by nodes-as-lines is particularly unique. As we have shown, simple visual estimates of proportions of classes of link groups provide intuition into abstract measures like S^3. Furthermore, as our Alignment Cycle layout demonstrates, it is possible to create new, specialized layout algorithms that exploit the linear ordering of nodes in ways that make it possible to quickly spot alignment pathologies.

Because of these features, VISNAB is a viable tool for studying objective functions and their performance. As discussed in [16], we strongly contend that the network alignment community should focus not on developing new network alignment algorithms but rather on devising more effective objective functions. Just as we experimented by combining S^3 and Importance, researchers should design custom objective functions for their specific needs to yield optimum results. To accomplish this task, we propose VISNAB as a tool researchers should utilize to help design objective functions.

For example, as we have shown in Case Study III, researchers can work toward creating objective functions that distribute edges and links into appropriate proportions of our identified groupings. As was shown in Case Study I, the sizes of these groupings can reflect the character of the alignment problem at hand, e.g., a pair of experiments where the larger network represented a deeper, more thorough investigation of the system, where new elements were added, and false positives were identified. So an objective function that worked to take this topology into account would be an improvement. Note that this approach can be useful *even in the absence of a gold standard alignment* for a given problem, as the overall proportions of the groupings could arguably apply to other alignments in the same class of problems. In other words, target vectors for the *NGS* and *LGS* scores could be estimated a priori for a class of problems.

4.4.1 Limitations

BioFabric's presentation of "nodes as lines" is unfamiliar for network researchers who have grown accustomed to visualizing nodes as points and can take some time getting used to. For networks small enough, the traditional node–link diagram approach is adequate, so it would be beneficial to allow the user to select a subset of the BioFabric network and view it using the traditional presentation. Going the other way, taking a subset of a traditional node–link presentation and viewing it using BioFabric, would be valuable as well. But this capability is not currently available.

Another shortcoming of the current VISNAB implementation is that the ordering of the node groups is fixed, per the order in Table 4.2. A different order may be more helpful in visualizing a particular alignment, so being able to dynamically reorder the node groups would be a useful feature.

Finally, just like the existing *NC* score, our new *NGS*, *LGS*, and *JS* scores depend upon the availability of a gold-standard alignment. Thus, they can really only be used to evaluate the performance of alignment algorithms and objective functions when the correct answer is already known. To be incorporated into a useful objective function, it would be necessary that for certain classes of alignment problems, target vectors for the *NGS* and *LGS* scores could be estimated a priori. It is also an open question about the difficulties of developing objective functions based upon incremental evaluation for these measures.

4.4.2 Future Work

An emerging area of investigation is the multiple network alignment problem, e.g., [24]. In these types of alignments, the fraction of the full set of networks that share a particular edge or node in the alignment is a crucial piece of information. This suggests that creating node and link groups for grouping elements with similar overlap percentages would be helpful in visualizing the quality and characteristics of a multiple network alignment.

Our new tool provides a platform to implement other new visualizations for exploring network alignments. Constructing and visualizing subgraphs of the full merged network (from **Network Merge**, above) can provide valuable insights as well. For example, one additional view that we support is the Orphan Edge layout, which shows the subgraph of G_1 that consists all the nodes connected by the orphan blue **pBp** edges, plus the first-degree neighbors of those nodes and corresponding connecting edges. In aligned networks where $E_1 \subseteq E_2$ (e.g., the *noisy yeast* network used in Case Study IV), blue edges will not exist at all in a correctly aligned network. Thus, viewing the context of the orphan blue **pBp** edges in the network can provide insights into alignment problems.

Similarly, we can visualize the common subgraph $CS_a = (V_1, E_a)$ of the alignment by displaying only the aligned purple **P** edges and the purple nodes they are incident upon. If this subgraph's nodes and edges are laid out using BioFabric's default layout [7], the various connected components of CS_a can be assessed. One way some researchers evaluate alignments is to determine if the alignment has a common subgraph with large connected regions [4, 15]. To this end, the user can visually estimate the components of the measure **Largest Common Connected Subgraph** (LCCS) using the heights of the nodes and widths of the edges of the connected components.

4.5 Conclusions

VISNAB provides a novel, powerful new way to visualize network alignments and will allow researchers to gain a new and better understanding of the strengths and shortcomings of the many available network alignment algorithms.

Appendix

Link and Node Groups with Blue Nodes

In the main text, for simplicity, we limited the discussion to the common case of aligning of network G_1 onto network G_2 when every node in G_1 is aligned onto a node in G_2. In the nomenclature we have introduced, this is an alignment "without any blue nodes." Tables 4.1 and 4.2 enumerated the possible link and node groups for these alignments. However, VISNAB is capable of handling alignments where unaligned blue nodes are permitted. In that case, the five link groups expand to seven, adding in groups three and four, which account for the case where blue edges are incident on blue nodes. Table 4.4 enumerates all possible link groups in the presence of blue nodes in the alignment.

In a similar fashion, the twenty node groups that can be present in an alignment without blue nodes expand to forty possible groups when blue nodes are allowed (Table 4.5). Note how, for example, Node Group 3 (see Table 4.2) splits into three distinct groups (3, 4, and 5) with blue nodes, since blue edges can be incident on

Table 4.4 Expansion of link groups when unaligned (blue) nodes are present

Link group	Edge color	Endpoint 1	Endpoint 2	Symbol
1	Purple	Purple	Purple	P
2	Blue	Purple	Purple	pBp
3	Blue	Purple	Blue	pBb
4	Blue	Blue	Blue	bBb
5	Red	Purple	Purple	pRp
6	Red	Purple	Red	pRr
7	Red	Red	Red	rRr

blue nodes as well as purple nodes. In a similar fashion, groups 7, 11, 14, 19, 23, 27, and 30 all split as well into three distinct groups. Groups 33–36 are introduced as well to account for blue nodes being present (Table 4.5).

Jaccard Similarity with Blue Nodes

Again, for simplicity, the main text only discussed the definition of our Jaccard Similarity (*JS*) score when there are no blue nodes present in the alignment. When blue nodes are not allowed, then for every node in G_1, we can find (using the correct alignment) where that node is supposed to go in G_2, and (using the given alignment) where it actually ends up in G_2. These two nodes in G_2 can then be compared to create the *JS* score for the node.

However, when blue nodes are allowed, there are four possible cases that can arise instead of one:

1. The node is supposed to be aligned, and it is (the case described above) ("purple node stays purple").
2. The node is supposed to be aligned, and it is not ("purple node turns to blue").
3. The node is not supposed to be aligned, and it is ("blue node turn to purple").
4. The node is not supposed to be aligned, and it is not ("blue node stays blue").

VISNAB handles case 4 by simply assigning a score of 1.0 to the node, since it is correctly left unaligned. To deal with cases 2 and 3, VISNAB instead compares two nodes in network G_1. Specifically, for case 2, if a node a in G_1 is supposed to (using the correct alignment) be aligned to node n in G_2 but is instead unaligned, we look to see which node b in G_1 is aligned (using the given alignment) to node n in G_2. We then create the *JS* score for node a by comparing the neighborhoods of a and b in G_1. If there is no node b (when nothing is aligned to node n in G_2, i.e., it is "red"), then the *JS* score for node a is 0.0. Case 3 is handled analogously, again comparing two nodes a and b in G_1 to obtain a *JS* score, with a 0.0 assigned if there is no matching node in G_1.

Table 4.5 Expansion of node groups when unaligned (blue) nodes are present

Node group	Node color	Incident edges	Symbol
1	Purple	None	(P:0)
2	Purple	Purple	(P:P)
3	Purple	pBp	(P:pBp)
4	Purple	pBb	(P:pBb)
5	Purple	pBp, pBb	(P:pBp/pBb)
6	Purple	pRp	(P:pRp)
7	Purple	Purple, pBp	(P:P/pBp)
8	Purple	Purple, pBb	(P:P/pBb)
9	Purple	Purple, pBp, pBb	(P:P/pBp/pBb)
10	Purple	Purple, pRp	(P:P/pRp)
11	Purple	pBp, pRp	(P:pBp/pRp)
12	Purple	pBb, pRp	(P:pBb/pRp)
13	Purple	pBp, pBb, pRp	(P:pBp/pBb/pRp)
14	Purple	Purple, pBp, pRp	(P:P/pBp/pRp)
15	Purple	Purple, pBb, pRp	(P:P/pBb/pRp)
16	Purple	Purple, pBp, pBb, pRp	(P:P/pBp/pBb/pRp)
17	Purple	pRr	(P:pRr)
18	Purple	Purple, pRr	(P:P/pRr)
19	Purple	pBp, pRr	(P:pBp/pRr)
20	Purple	pBb, pRr	(P:pBb/pRr)
21	Purple	pBp, pBb, pRr	(P:pBp/pBb/pRr)
22	Purple	pRp, pRr	(P:pRp/pRr)
23	Purple	Purple, pBp, pRr	(P:P/pBp/pRr)
24	Purple	Purple, pBb, pRr	(P:P/pBb/pRr)
25	Purple	Purple, pBp, pBb, pRr	(P:P/pBp/pBb/pRr)
26	Purple	Purple, pRp, pRr	(P:P/pRp/pRr)
27	Purple	pBp, pRp, pRr	(P:pBp/pRp/pRr)
28	Purple	pBb, pRp, pRr	(P:pBb/pRp/pRr)
29	Purple	Blue, pRp, pRr	(P:pBp/pBb/pRp/pRr)
30	Purple	Purple, pBp, pRp, pRr	(P:P/pBp/pRp/pRr)
31	Purple	Purple, pBb, pRp, pRr	(P:P/pBb/pRp/pRr)
32	Purple	Purple, pBp, pBb, pRp, pRr	(P:P/pBp/pBb/pRp/pRr)
33	Blue	pBb	(B:pBb)
34	Blue	bBb	(B:bBb)
35	Blue	pBb, bBb	(B:pBb/bBb)
36	Blue	None	(B:0)
37	Red	pRr	(R:pRr)
38	Red	rRr	(R:rRr)
39	Red	pRr, rRr	(R:pRr/rRr)
40	Red	None	(R:0)

For some network $G = (V, E)$, let $N_G(z_1) = \{z_2 \in V : (z_1, z_2) \in E\}$ be the neighborhood of node z_1 in G. For nodes $x, y \in V$, let $N_G(x, y)$ be the neighborhood of x disregarding y, and let i_{xy} be a corrective term accounting for a possible edge between the two. Accordingly, if $y \in N_G(x)$, then $N_G(x, y) = N_G(x) - y$ and $i_{xy} = 1$, else $N_G(x, y) = N_G(x)$ and $i_{xy} = 0$. Let $N_G(y, x)$ be defined analogously. Our extended *JS* definition $\sigma_G : V \times V \to [0, 1]$ between two nodes is defined as

$$\sigma_G(x, y) = \frac{|N_G(x, y) \cap N_G(y, x)| + i_{xy}}{|N_G(x, y) \cup N_G(y, x)| + i_{xy}}.$$

Note that when x and y are both singletons, we define $\sigma(x, y) = 1.0$ to avoid dividing by zero.

$$f_a(u) = \begin{cases} \sigma_{G_2}(a(u), a_c(u)) & \text{if } a(u) \text{ and } a_c(u) \text{ are defined} \\ \sigma_{G_1}(u, a_c^{-1}(a(u))) & \text{if } a(u) \text{ is defined} \\ \sigma_{G_1}(u, a^{-1}(a_c(u))) & \text{if } a_c(u) \text{ is defined} \\ 1.0 & \text{if } a(u) \text{ and } a_c(u) \text{ are undefined} \end{cases}$$

Given node sets $V_a, V_c \in V_1$, an alignment $a : V_a \to V_2$, and the correct alignment $a_c : V_c \to V_2$, our *JS* measure for the given alignment a, with respect to the correct alignment a_c, is defined as

$$JS(a) = \frac{1}{|V_1|} \sum_{u \in V_1} f_a(u).$$

If the correct alignment is provided, the user can choose to have correctly and incorrectly aligned nodes laid out separately in different node groups. The user can choose the criterion for the correct alignment to be based either on the traditional *NC* measure or on our *JS* measure. If *JS* is chosen, the user can set the threshold value $\beta \in [0, 1]$, so, a node in the form of u::v or u:: is denoted correct if $f_a(u) \geq \beta$.

Creation of the Correct Network Alignment

To create the "correct" alignment used in the case studies, we wanted to create two networks where all nodes in the smaller network had one and only one known matching node in the larger network. One consequence of this approach is that our correct alignment did not have any blue nodes. These case studies use two different protein–protein interaction datasets. The larger "SC" network, from *S. cerevisiae*, contains 5831 nodes and 77,149 edges and was originally obtained from BioGRID (v3.2.101, June 2013) [14]. The smaller "Yeast2" network, also from *S. cerevisiae*, has 2390 nodes and 16,127 edges. It was originally generated from data in [13] and used in [4]. Both networks were previously used in [16].

Nodes in Yeast2 are tagged with a variety of gene symbols (e.g., *PSY4*), secondary identifiers, and synonyms, while nodes in SC were tagged with ENTREZ IDs (e.g., 852234). In order to generate the "correct" alignment file, it was necessary to find the mapping from the former to the latter. To do this, we first used the YeastMine API [25, 26] at https://yeastmine.yeastgenome.org, provided by the Saccharomyces Genome Database (SGD) [23], in order to generate a mapping from the node names to the SGD IDs that we could then feed to the DAVID web tool [27, 28]. With a Java program employing libraries provided by `org.intermine`, we downloaded (11 Feb. 2018) tuples for `Gene.primaryIdentifier`, `Gene.secondaryIdentifier`, `Gene.symbol`, and `Gene.synonyms.value` for `Gene.organism.shortName="S. cerevisiae"`, for all entries in the lists `Verified_ORFs`, `Dubious_ORFs` and `ALL_Verified_Uncharacterized_Dubious_ORFs`. Three remaining genes *YAR010C*, *YBR012W-B*, and *YHL009W-B* were not in any of these lists and were explicitly queried.

For each gene in Yeast2, we then matched the node name to a `Gene.synonyms.value`, and from this obtained a list of one or more `Gene.primaryIdentifiers`. In the cases where there was more than one, we chose the `Gene.primaryIdentifier` that mapped to a Gene.symbol that matched the `Gene.synonyms.value`. For example, synonym *MSL1* mapped to SGD IDs S000004374 and S000001448. However, while the former SGD ID mapped to gene symbol *NAM2*, the latter mapped to *MSL1* and thus was selected. With one exception, this approach resulted in an unambiguous mapping of all Yeast2 node names to SGD IDs. The exception was for gene names *EFG1* and *YGR272C*; the latter was merged into the former, giving both names the same SGD ID (S000007608). Thus, node *YGR272C* was dropped (see Table 4.6).

These SGD IDs were then uploaded as a gene list to DAVID at:

https://david.ncifcrf.gov/conversion.jsp

(DAVID 6.8, accessed 18 Feb. 2018). Since DAVID has restrictions on large-scale queries through their web API, this was done manually. Upon uploading the list, DAVID's Gene List Manager was not able to identify five IDs, so these nodes were dropped as well (see Table 4.7).

We instructed the tool to convert SGD_IDs to ENTREZ_GENE_IDs and downloaded the result. Thus, we had a mapping of 2384 of the nodes in Yeast2 to ENTREZ IDs. However, not all of these ENTREZ IDs are present as nodes in the larger SC network. In order to create a correct alignment with no blue nodes, we then pruned the Yeast2 network to remove the small number of nodes that could

Table 4.6 Node that was merged and dropped

Gene name	SGD ID
YGR272C	S000007608

Table 4.7 Nodes not identified by DAVID and dropped

Gene name	SGD ID
IMD1	S000000095
YNL276C	S000005220
YDR133C	S000002540
YDL026W	S000002184
YAR075W	S000002145

Table 4.8 Nodes that could not be mapped to SC network and dropped

Gene name	SGD ID
ATM1	855347
PHM8	856759
PUT1	850833
CTM1	856509
SBE2	851953

not be mapped onto the SC network. This resulted in an additional five nodes that needed to be dropped; see Table 4.8.

Detailed Description of the Node Assignment Algorithm for the Node and Link Group Layout

The node assignment algorithm for the Node and Link Group Layout is a multiqueue breadth-first search graph traversal. While a typical breadth-first search utilizes a single queue, our multiqueue approach uses one queue for each node group, and the queues are processed in the order listed in Table 4.2. The traversal starts on the node of highest degree in the first queue; its neighbor nodes are then visited in the order of decreasing degree. If a newly visited node is in the current node group, it will be placed onto the current queue; if it is not, it will be placed onto the queue of its node group. The traversal is finished with a queue when every node in that node group has been visited. If the queue is empty but there still are unvisited nodes in that group, the highest degree node from the set of unvisited nodes of that group is added to the queue; after the queue is traversed, if there still are unvisited nodes in the group, this step is repeated until all nodes in the group are visited. Once finished, the traversal moves to the next queue. If a queue is empty when first evaluated, the node of highest degree in that queue's node group is added.

Thus, we created a "Yeast2-reduced" network consisting of 2379 nodes and 16,063 edges, which was used in the case studies.

Table 4.9 All alignment scores for Case Study III

Alignment	NGS	LGS	NC	JS	S^3	Resnik
Correct	1.00	1.00	1.00	1.00	0.25	9.63
$1.0 * I$	0.61	0.79	0.00042	0.021	0.0043	3.16
$0.001 * S^3 + 0.999 * I$	0.88	0.86	0.00042	0.024	0.17	3.48
$0.003 * S^3 + 0.997 * I$	0.88	0.86	0.00042	0.021	0.18	3.39
$0.005 * S^3 + 0.995 * I$	0.72	0.85	0.00	0.025	0.10	3.27
$0.01 * S^3 + 0.99 * I$	0.88	0.86	0.00	0.024	0.19	3.32
$0.03 * S^3 + 0.97 * I$	0.87	0.80	0.018	0.057	0.27	3.62
$0.05 * S^3 + 0.95 * I$	0.73	0.53	0.022	0.063	0.49	3.44
$0.1 * S^3 + 0.9 * I$	0.67	0.48	0.017	0.067	0.54	3.50
$1.0 * S^3$	0.64	0.46	0.021	0.069	0.55	3.61

Full Table of All Alignment Scores for Mixtures of Importance and Symmetric Substructure Score

Table 4.9 lists the scores for the 10-h SANA [16] runs between Yeast2K-Reduced and SC, in which we used combinations of Importance (I) [17] and Symmetric Substructure Score (S^3) [15] in the objective function. Note that all these scores, with the exception of Resnik, are available using the Alignment Measures tool in VISNAB. The Resnik scores [18–20] shown here are the means of the nonzero, non-"None" values computed separately using FastSemSim [21], incorporating Gene Ontology (GO) terms [29, 30] downloaded in February 2019.

Table of Node Group Sizes for Case III

Table 4.10 provides the number of nodes in each node group for the four alignments discussed in Case III. Asterisks show the four largest node groups per alignment, which are labeled prominently in Fig. 4.5. As called out in the text, of the 716 nodes in the top 13 rows above group **(P:P/pRp/pRr)** for the mixed alignment, 544 (76%) have no incident **pRr** edges.

Percentage of Purple Nodes Without and with Incident pRr Edges Between Correct and Mixed Alignments

The discussion of Fig. 4.5 notes that while there are *more* **pRr** edges in the mixed alignment compared to the correct alignment, those edges are concentrated across a *smaller fraction* of the purple nodes in that mixed alignment. Table 4.11 compares

Table 4.10 Sizes of all node groups in Case III

Symbol	Correct	All importance	Mixed	All S^3
(P:0)	0	0	0	0
(P:P)	2	0	23	16
(P:pBp)	0	0	0	0
(P:pRp)	0	0	0	0
(P:P/pBp)	2	0	59	23
(P:P/pRp)	52	0	62	3
(P:pBp/pRp)	32	403*	207	0
(P:P/pBp/pRp)	27	9	193	1
(P:pRr)	0	0	0	0
(P:P/pRr)	5	0	31	439
(P:pBp/pRr)	5	35	98	303
(P:pRp/pRr)	0	0	0	0
(P:P/pBp/pRr)	1	9	43	662*
(P:P/pRp/pRr)	981*	0	530*	157
(P:pBp/pRp/pRr)	310	1677*	82	0
(P:P/pBp/pRp/pRr)	962*	246	1051*	775*
(R:pRr)	578*	843*	752 *	31
(R:rRr)	209	58	0	1087*
(R:pRr/rRr)	2665*	2551*	2700*	2334*
(R:0)	0	0	0	0

Table 4.11 Comparison of purple node fractions without and with **pRr** edges

Alignment	(P:*)	(P:*/pRr)
Correct alignment	4.83%	95.17%
Mixed alignment	22.87%	77.13%

the percentages of all purple nodes without [**(P:*)**] and with [**(P:*/pRr)**] **pRr** incident edges, between the correct and mixed alignments.

Alignment Cycle Layout with Blue Nodes

When unaligned blue nodes are not allowed, there are four cases that must be handled by the Alignment Cycle layout, and these are indicated by a checkmark in the rightmost column of Table 4.12. When unaligned blue nodes are present, there are nine path types that must be handled. In the table, a network with nodes $\{A, B, C, \ldots L\}$ has been aligned onto a network with nodes $\{1, 2, 3, \ldots 12\}$. Note that purple node runs can extend for any length of nodes, as shown by the $\cdots$, but the matches and alignments given in this table are for the cases where there are none of these extra nodes. The Alignment Cycle layout will order the nodes in the path and cycle cases so that misaligned nodes are laid out next to their correct partners; see case 9 in particular to see this pattern.

Table 4.12 Alignment cycle layout cases

#	Layout	Correct alignment	Test alignment	Type	No blue
1	(A::)	$A \to \emptyset$	$A \to \emptyset$	Correct	
2	(::1)	$\emptyset \to 1$	$\emptyset \to 1$	Correct	✓
3	(B::2)	$B \leftrightarrow 2$	$B \to 2$	Correct	✓
4	(C::3)	$C \to \emptyset$	$C \to 3$	Path	
		$\emptyset \to 3$			
5	(::4) (D::)	$4 \leftrightarrow D$	$\emptyset \to 4$	Path	
			$D \to \emptyset$		
6	(::5) (E::6) ···	$5 \leftrightarrow E$	$\emptyset \to 5$	Path	✓
		$\emptyset \to 6$	$E \to 6$		
7	(F::7) ··· (G::)	$7 \leftrightarrow G$	$F \to 7$	Path	
		$F \to \emptyset$	$G \to \emptyset$		
8	(::8) (H::9) ··· (I::)	$8 \leftrightarrow H$	$\emptyset \to 8$	Path	
		$9 \leftrightarrow I$	$H \to 9$		
			$I \to \emptyset$		
9	(J::10) (K::11) ··· (L::12)	$10 \leftrightarrow K$	$J \to 10$	Cycle	✓
		$11 \leftrightarrow L$	$K \to 11$		
		$12 \leftrightarrow J$	$L \to 12$		

The Four-Cluster Misalignment

In Case Study IV, we showed how the Alignment Cycle layout could be used to spot alignment problems such as two entire protein clusters being swapped. Figure 4.8 shows a severe degeneracy for the same alignment run, where *four* separate protein clusters were misaligned in a cycle. The BioFabric depiction of this problem follows the same pattern shown in Fig. 4.7, but the successive edge wedges are even steeper here and show a clear pattern of cycling between four distinct sets of node rows.

Availability and Requirements

- **Project Name** VISNAB: Visualization of Network Alignments using BioFabric
- **Project Home Page** http://www.BioFabric.org/VISNAB/index.html. The VISNAB code repository is at https://github.com/wjrl/AlignmentPlugin, while the BioFabric code repository is at https://github.com/wjrl/BioFabric.
- **Operating Systems** Cross-platform.
- **Programming Language** Java
- **Other Requirements** Plugin requires BioFabric Version 2 Beta Release 2 or above, available on GitHub and the BioFabric project Home Page. An OpenJDK Java runtime is bundled with BioFabric and does not need to be installed separately.
- **License** LGPL V 2.1. Per the LGPL license, the VISNAB source code is provided in additional file 2.

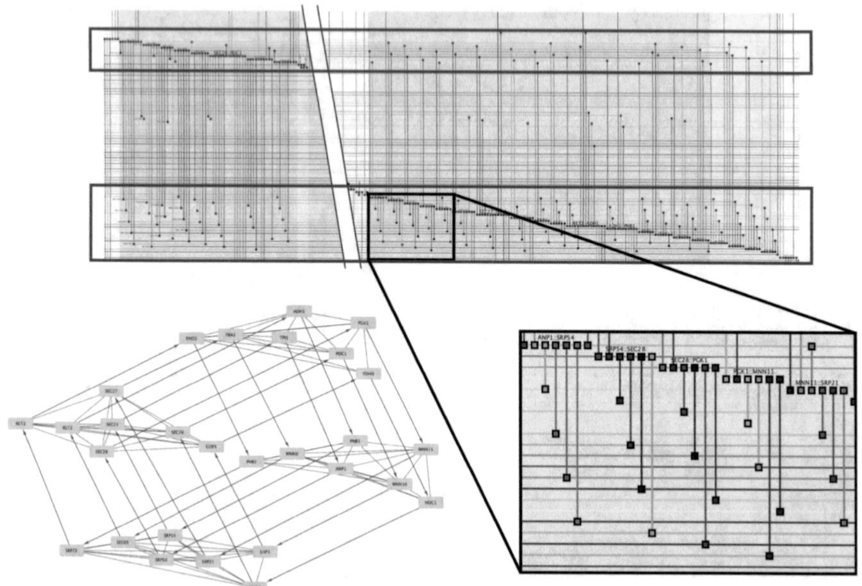

Fig. 4.8 An even more striking misalignment, where four different protein complexes have been swapped in a round-robin fashion. The traditional node–link diagram is shown at the lower left with edges (colored blue) for the protein–protein interactions and directed edges (colored red) for the alignments. The four protein complexes clockwise from top (per SGD): (1) glycolysis- and gluconeogenesis-related genes, (2) mannosyltransferase complex and prohibitin complex, (3) signal recognition particle, and (4) the coatomer complex (COPI). The BioFabric layout on the top, shown in detail at the lower right, shows the distinct pattern displayed by this artifact, with adjacent edge wedges having edges cycling every fourth node. Note that a slice was removed from the upper view because the three separate cycles constituting this structure are not contiguous in the layout

List of Abbreviations

EC	Edge Coverage, Conservation, Correctness, or Correspondence
ICS	Induced Conserved Structure
JS	Jaccard Similarity
LCCS	Largest Common Connected Subgraph
LGS	Link Group Similarity
NC	Node Correctness
NGS	Node Group Similarity
PPI	protein–protein interaction
(S^3)	Symmetric Substructure Score
SGD	Saccharomyces Genome Database
VISNAB	Visualization of Network Alignments using BioFabric

Additional Files

Archived executables are available at: https://www.ics.uci.edu/~wayne/research/papers/BioFabricVISNAB.

- Additional file 1: DesaiEtAl-2020-CaseStudyFiles.zip
  Archive of .sif, .gw, .align, .resnik, and .bif (BioFabric) files for the four case studies. The latter can be loaded into BioFabric with VISNAB to view the case study results in detail.
- Additional file 2: VISNAB-1.1.0.0-src.tar.gz
  Contains the source code for the VISNAB V1.1 plugin. The most recent source code is available at the GitHub repository listed above.
- Additional file 3: BioFabric-2.0.B.2-src.tar.gz
  Contains the source code for the beta version of BioFabric 2.0 needed to run the plugin. The most recent source code is available at the GitHub repository listed above.
- Additional file 4: sVISNAB-V1.1.0.0.jar
  The plugin jar for VISNAB V1.1.
- Additional file 5: BioFabric-2.0.B.2.tgz
  Contains a Linux executable for a beta version of BioFabric 2.0 needed to run the plugin. Executables for Windows and Mac are available on GitHub. In a pinch, the bundled OpenJDK for Linux in this provided executable can be swapped out for other versions to run the program on Windows or Mac from a batch file or shell script.

References

1. Milano, M., Guzzi, P.H., Tymofieva, O., Xu, D., Hess, C., Veltri, P., et al.: Multiple anonymized social networks alignment. In: 2015 IEEE International Conference on Data Mining, pp. 599–608 (2015)
2. Milano, M., Tymofiyeva, O., Xu, D., Hess, C., Cannataro, M., Guzzi, P: Using network alignment for analysis of connectomes: experiences from a clinical dataset. In: Proceedings of the 7th ACM International Conference on Bioinformatics, Computational Biology, and Health Informatics. BCB'16, pp. 649–656. ACM, New York (2016). Available from: https://dl.acm.org/citation.cfm?doid=2975167.2985690
3. Milano, M., Guzzi, P., Tymofiyeva, O., Xu, D., Hess, C., Veltri, P., et al.: An extensive assessment of network alignment algorithms for comparison of brain connectomes. BMC Bioinf. **18**, 235 (2017)
4. Kuchaiev, O., Milenković, T., Memišević, V., Hayes, W., Pržulj, N.: Topological network alignment uncovers biological function and phylogeny. J. R. Soc. Interface **7**(50), 1341–1354 (2010)
5. Uetz, P., Dong, Y.A., Zeretzke, C., Atzler, C., Baiker, A., Berger, B., et al.: Herpesviral protein networks and their interaction with the human proteome. Science **311**(5758), 239–242 (2006)
6. Milenković, T., Zhao, H., Faisal, F.E.: Global network alignment in the context of aging. In: Proceedings of the International Conference on Bioinformatics, Computational Biology and Biomedical Informatics. BCB'13, pp. 23:23–23:32. ACM, New York (2013). Available from: https://doi.org/10.1145/2506583.2508968
7. Longabaugh, W.J.R.: Combing the hairball with BioFabric: a new approach for visualization of large networks. BMC Bioinf. **13**, 275 (2012)
8. Tamassia, R., Tollis, I.G.: A unified approach to visibility representations of planar graphs. Discrete Comput. Geom. **1**(4), 321–341 (1986)

9. Blakley, B.: Reduction of flow diagrams to unfolded form modulo snarls. Defense Technical Information Center; 1987. To AFOSR on Contract F49620-86-C-0103
10. McAllister, A.J.: A new heuristic algorithm for the linear arrangement problem. Faculty of Computer Science, University of New Brunswick, 1999. 99_126a
11. Xie, J., Xiang, C., Zhou, Z., Dai, D., Zhang, H.: NetCompare: a visualization tool for network alignment on Galaxy. In: 2014 International Conference on Information Science, Electronics and Electrical Engineering, vol. 2, pp. 881–884 (2014)
12. Malek, M., Ibragimov, R., Albrecht, M., Baumbach, J.: CytoGEDEVO-global alignment of biological networks with Cytoscape. Bioinformatics **32**(8), 1259–1261 (2016)
13. Collins, S.R., Kemmeren, P., Zhao, X.C., Greenblatt, J.F., Spencer, F., Holstege, F.C.P., et al.: Toward a comprehensive atlas of the physical interactome of Saccharomyces cerevisiae. Mol. Cell. Proteomics **6**(3), 439–450 (2007). Available from: http://www.mcponline.org/content/6/3/439.abstract
14. Chatr-aryamontri, A., Breitkreutz, B.J., Heinicke, S., Boucher, L., Winter, A., Stark, C., et al.: The BioGRID interaction database: 2013 update. Nucleic Acids Res. **41**(D1), D816–D823 (2013) Available from: http://nar.oxfordjournals.org/content/41/D1/D816.abstract
15. Saraph, V., Milenković, T.: MAGNA: maximizing accuracy in global network alignment. Bioinformatics **30**(20), 2931–2940 (2014)
16. Mamano, N., Hayes, W.B.: SANA: simulated annealing far outperforms many other search algorithms for biological network alignment. Bioinformatics **33**(14), 2156–2164 (2017)
17. Hashemifar, S., Xu, J.: HubAlign: an accurate and efficient method for global alignment of protein–protein interaction networks. Bioinformatics **30**(17), i438–i444 (2014) Available from: https://doi.org/10.1093/bioinformatics/btu450
18. Resnik, P.: Using information content to evaluate semantic similarity in a taxonomy. In: Proceedings of the 14th International Joint Conference on Artificial Intelligence - Volume 1. IJCAI'95, pp. 448–453. Morgan Kaufmann Publishers Inc., San Francisco (1995). Available from: http://dl.acm.org/citation.cfm?id=1625855.1625914
19. Lord, P.W., Stevens, R.D., Brass, A., Goble, C.A.: Semantic similarity measures as tools for exploring the gene ontology. In: Proceedings of the 8th Pacific Symposium on Biocomputing, pp. 601–612 (2003)
20. Lord, P.W., Stevens, R.D., Brass, A., Goble, C.A.: Investigating semantic similarity measures across the Gene Ontology: the relationship between sequence and annotation. Bioinformatics **19**(10), 1275–1283 (2003). Available from: https://doi.org/10.1093/bioinformatics/btg153
21. Guzzi, P.H., Mina, M., Guerra, C., Cannataro, M.: Semantic similarity analysis of protein data: assessment with biological features and issues. Briefings Bioinf. **13**(5), 569–585 (2012)
22. Shannon, P., Markiel, A., Ozier, O., Baliga, N.S., Wang, J.T., Ramage, D., et al.: Cytoscape: a software environment for integrated models of biomolecular interaction networks. Genome Res. **13**(11), 2498–2504 (2003)
23. Cherry, J., Adler, C., Ball, C., Chervitz, S.A., Dwight, S., Hester, E., et al.: SGD: Saccharomyces genome database. Nucleic Acids Res. **26**(1), 73–79 (1998)
24. Vijayan, V., Milenković, T.: Multiple network alignment via multiMAGNA++. IEEE/ACM Trans. Comput. Biol. Bioinf. **15**, 1669–1682 (2018)
25. Balakrishnan, R., Park, J., Karra, K., Hitz, B.C., Binkley, G., Hong, E.L., et al.: YeastMine—an integrated data warehouse for Saccharomyces cerevisiae data as a multipurpose tool-kit. Database **2012**, bar062 (2012). Available from: https://doi.org/10.1093/database/bar062
26. Smith, R.N., Aleksic, J., Butano, D., Carr, A., Contrino, S., Hu, F., et al.: InterMine: a flexible data warehouse system for the integration and analysis of heterogeneous biological data. Bioinformatics **28**(23), 3163–3165 (2012). Available from: https://doi.org/10.1093/bioinformatics/bts577
27. Huang, D.W., Sherman, B.T., Lempicki, R.A.: Bioinformatics enrichment tools: paths toward the comprehensive functional analysis of large gene lists. Nucleic Acids Res. **37**(1), 1–13 (2009). Gkn923[PII],19033363[pmid]. Available from: http://www.ncbi.nlm.nih.gov/pmc/articles/PMC2615629/

28. Huang, D.W., Sherman, B.T., Lempicki, R.A.: Systematic and integrative analysis of large gene lists using DAVID bioinformatics resources. Nat. Protoc. **4**, 44 EP (2008). Available from: https://doi.org/10.1038/nprot.2008.211
29. Ashburner, M., Ball, C.A., Blake, J.A., Botstein, D., Butler, H., Cherry, J.M., et al.: Gene ontology: tool for the unification of biology. The Gene Ontology Consortium. Nat Genet. **25**, 25–29 (2000)
30. Consortium TGO: The gene ontology resource: 20 years and still going strong. Nucleic Acids Res. **47**, D330–D338 (2019)

Part II
Network Module Detection

Chapter 5
Motifs in Biological Networks

Rasha Elhesha, Aisharjya Sarkar, and Tamer Kahveci

Abstract Biological networks provide great potential to understand how cells function. Motifs in biological networks, frequent topological patterns, represent key structures through which biological networks operate. Studying motifs answers important biological questions. Finding motifs in biological networks remains to be a computationally challenging task as the sizes of the motif and the underlying network grow. Several algorithms exist in the literature to solve this problem. This chapter discusses the biological significance of network motifs, motivation behind solving the motif detection problem and the key challenges of this problem. We discuss different formulations of motif detection problem based on several orthogonal perspectives that change the problem definition as well as solution significantly. The first perspective considers the number of input networks involved (i.e., one or more than one networks). The second perspective focuses on the labeling (i.e., labeled or unlabeled) of the nodes and edges of the input network. The third one considers different frequency definitions of counting motif instances (i.e., F1, F2, and F3) in a network. The fourth perspective describes whether the underlying network is directed or undirected. The last one considers motif detection under different types of network models (i.e., deterministic, probabilistic, or dynamic model). As a case study for each formulation, we briefly discuss important existing methods from the literature. Finally, we conclude with future research directions.

Keywords Biological network · Network motifs · Maximum independent set · Motif frequency · Subgraph isomorphism

---

R. Elhesha · A. Sarkar · T. Kahveci (✉)
University of Florida, Gainesville, FL, USA
e-mail: tamer@cise.ufl.edu

© Springer Nature Switzerland AG 2021

B.-J. Yoon, X. Qian (eds.), *Recent Advances in Biological Network Analysis*,
https://doi.org/10.1007/978-3-030-57173-3_5

5.1 Introduction and Background

Biological networks describe how molecules interact to carry out various cellular functions. Protein–protein interaction (PPI), gene regulation network (GRN) are two such examples of biological networks. One common way to represent biological networks is to use graphs, where the nodes and the edges represent the interacting molecules and the interactions between those molecules, respectively, [57]. Studying biological networks has great potential to understand how cells function and how they respond to extracellular stimulants. Such studies have already been used successfully in many applications. For instance, characterizing the variations in drug resistance of different cell lines [12], identifying the pathways serving similar functions across different organisms [5, 51] are two such applications among many others.

Motifs are frequent topological patterns within a given network [35]. Identifying motifs has been one of the key steps in understanding the functions served by various types of biological networks such as GRN or PPI networks [2, 44, 49, 52]. Motifs help to uncover the basic structure and design principles of a network [32]. They are often considered as the basic building blocks of a network [35], and even one of the local properties of a network [34]. Thus, they can be used to classify networks [16] into functional sub-units. It is worth noting that motifs have been used in various applications such as validating the evolutionary mechanism of phylogeny using parsimony models [38] and prediction of regulatory elements in genomic sequences [54].

Although studying motifs is of utmost importance for network analysis, motif identification remains to be a computationally hard problem [22]. The challenges arise due to several reasons. First, even when the motif topology is given, counting motif frequency (i.e., the number of occurrences of this motif) requires solving the subgraph isomorphism problem, which is NP-complete [15]. Second, when the motif topology is not known in advance, trying out all possible alternative topologies is not feasible since the number of such topologies increases exponentially with the number of edges in the motif.

Motif counting problem becomes computationally harder for biological networks as biological interactions are inherently stochastic events [9]. The interactions within a network may or may not take place depending on the outcome of various biological events [45]. For example, a disease or a viral infection in an organism may lead to genetic abnormalities which alternate transcription, RNA composition, or protein structure. This in turn may affect the likelihood of interactions between molecules. In addition, timing and location of the biological events may yield differences in network topology as well. For example, DNA replication may initiate from varying locations at different times of the cell cycle that can alter the expression levels of the genes [20, 43], thereby generating different GRN topologies.

Various existing studies model the uncertainty in biological interactions using probabilistic graphs. Briefly, this model assigns a probability value for each edge of the graph that represents the probability of observing that interaction [45]. If

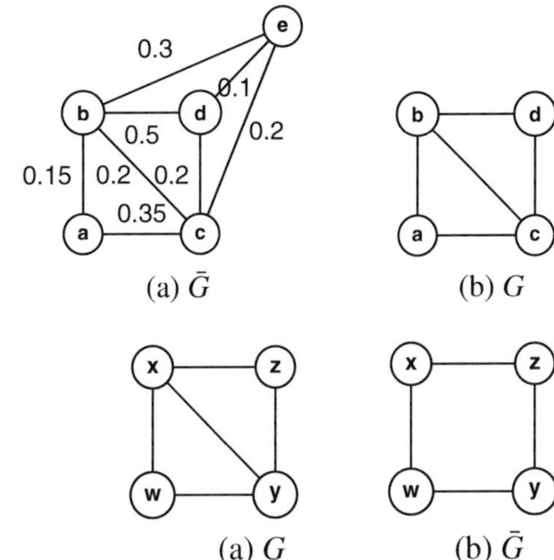

Fig. 5.1 (**a**) A probabilistic network with five nodes and eight edges. The numbers on each edge denote the existence probability of that edge. (**b**) A possible deterministic instance of the probabilistic network in (**a**)

Fig. 5.2 (**a**) A hypothetical network with four nodes and five edges. (**b**) A possible dynamic state of the network in (**a**) due to absence of the edge (x, y)

a biological network contains at least one probabilistic interaction, we call it a *probabilistic network*. Otherwise, we call it a *deterministic network*. Figure 5.1a depicts a hypothetical probabilistic network that has five nodes and eight edges. The number on each edge represents the probability of occurrence of the corresponding interaction. Probabilistic network is a summary of all possible deterministic networks formed by different subsets of interactions. More specifically, a probabilistic network, represented as a graph with $|E|$ edges, summarizes $2^{|E|}$ deterministic networks, which may arise as instances of the probabilistic network. Figure 5.1b presents one such instance of the probabilistic network in Fig. 5.1a.

In several types of biological networks, the activity levels of the interactions between molecules can change over time [4]. For instance, when a metabolic network reaches from a transient to a steady state, a subset of its fluxes gradually change. The sequence of those intermediate states is called the *dynamic* state of the given biological network. Analysis of such dynamic states, captured through a sequence of graphs, is crucial as those states may lead to undesired effects such as toxicity. Such dynamic changes in topology of a biological network with time are often modeled through network edit operations. These edit operations are represented as insertion and deletion of interactions (edges) between molecules [6]. Figure 5.2b represents a hypothetical dynamic state of the biological network in Fig. 5.2a due to one edge deletion (x, y).

The rest of this chapter is organized as follows. We present the key definitions necessary to understand our topic in Sect. 5.2. Next, we describe different existing motif discovery algorithms in Sect. 5.3. Finally, we end with a brief conclusion in Sect. 5.4.

5.2 Definitions and Notation

We represent a given biological network using a graph denoted with $G = (V, E)$. Here, the set of nodes V denotes the set of molecules, and the set of edges E denotes the interactions among those molecules. In the rest of this chapter, we use the term graph to denote a biological network.

In a graph G, the set E may represent *directed* or *undirected* edges depending on the type of biological network represented by G [11]. Figure 5.3 shows two networks one undirected and the other directed. An example to biological networks (graph) with directed interactions (edges) is a transcriptional regulatory network where nodes represent transcription factors and genes, while edges represent transcriptional regulation [31]. In such a network, a transcription factor (node) acts as the source of a direct regulatory edge to a target gene (another node) by producing RNA or protein molecule that functions as transcriptional activator or inhibitor thereby regulating the target gene. An example of undirected biological network is PPI where proteins are nodes and edges are interactions between those proteins [28].

The sets of nodes V or edges E in a graph G can be *labeled* or *unlabeled.* A label represents the name or identity for an entity (i.e., edge or node) [8]. Notice that the labels may not be unique. Figure 5.4 shows four different scenarios of a hypothetical labeled/unlabeled graph, namely edge-labeled, node-labeled, edge- and node-labeled, and unlabeled graphs. Several studies include labels of nodes and edges in their work to have coherent results in terms of the interactions or the interacting molecules. For example, finding labeled network motifs in a labeled PPI network can be used to predict unknown protein function.

We say that a graph is *connected* if there exists a path between all pairs of its nodes in V. We say that a graph $S = (V_S, E_S)$ is a *subgraph* of G if $V_S \subseteq V$ and $E_S \subseteq E$. In the rest of this chapter, we only consider connected subgraphs. In the rest of this chapter, we use the term subgraph instead of connected subgraph to

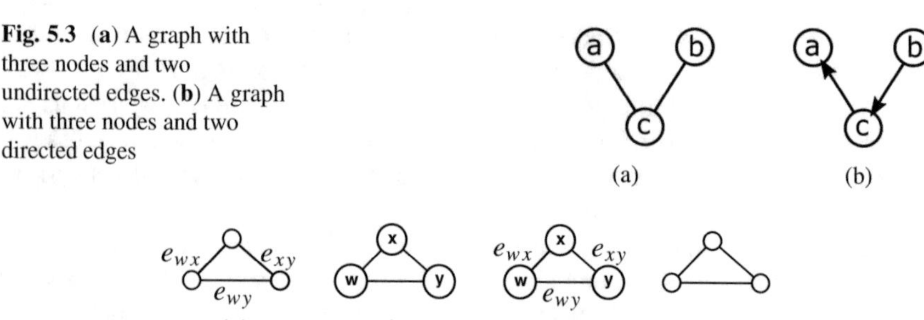

Fig. 5.3 (**a**) A graph with three nodes and two undirected edges. (**b**) A graph with three nodes and two directed edges

Fig. 5.4 (**a**) A hypothetical edge-labeled graph with three nodes and three edges. (**b**) A hypothetical node-labeled graph. (**c**) A hypothetical node and edge-labeled graph. (**d**) A hypothetical unlabeled graph

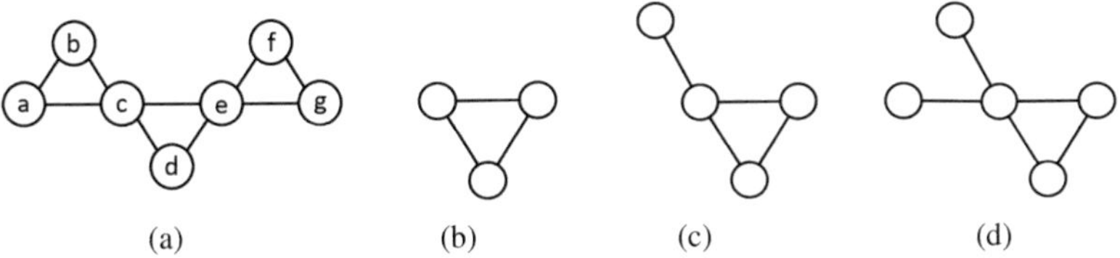

Fig. 5.5 (**a**) A graph G that contain seven nodes {a, b, c, d, e, f, g} and nine edges {(a,b), (a,c), (b,c), (c,d), (c,e), (d,e), (e,f), (e,g), (f,g)}. (**b**) A pattern with three embeddings in G, {(a,b), (a,c), (b,c)}, {(c,d), (c,e), (d,e)} and {(e,f), (e,g), (f,g)}. (**c**) A pattern with eight embeddings in G, S_1 = {(a,b), (a,c), (b,c), (c,d)}, S_2 = {(a,b), (a,c), (b,c), (c,e)}, S_3 = {(e,f), (e,g), (f,g), (c,e)}, S_4 = {(e,f), (e,g), (f,g), (d,e)}, S_5 = {(c,d), (c,e), (d,e), (a,c)}, S_6 = {(c,d), (c,e), (d,e), (b,c)}, S_7 = {(c,d), (c,e), (d,e), (e,f)} and S_8 = {(c,d), (c,e), (d,e), (e,g)}. (**d**) A pattern that has four copies in G, S_9 = {(a,b), (a,c), (b,c), (c,d), (c,e)}, S_{10} = {(e,f), (e,g), (f,g), (c,e), (d,e)}, S_{11} = {(c,d), (c,e), (d,e), (a,c), (b,c)} and S_{12} = {(c,d), (c,e), (d,e), (e,f), (e,g)}

simplify our terminology, unless otherwise specified. Notice that a subgraph of a given graph can be uniquely determined by the set of edges E_S of that subgraph as all of its nodes are connected. Given G, we say that two subgraphs $S_1 = (V_{S_1}, E_{S_1})$ and $S_2 = (V_{S_2}, E_{S_2})$ are *identical* if they have the same set of edges. A less constrained association between two subgraphs is *isomorphism*. Two subgraphs S_1 and S_2 are isomorphic if the following condition holds: There exists a bijection $f : V_{S_1} \rightarrow V_{S_2}$ such that $\forall (u, v) \in E_{S_1} \iff (f(u), f(v)) \in E_{S_2}$. We say that two subgraphs S_1 and S_2 *overlap* if they share at least one edge (i.e., $E_{S_1} \cap E_{S_2} \neq \emptyset$). In Fig. 5.5a, consider the three subgraphs S_1, S_2, and S_3 defined by the set of edges {(a,b), (a,c), (b,c), (c,d)}, {(a,b), (a,c), (b,c), (c,e)}, and {(e,f), (e,g), (f,g), (c,e)}, respectively. S_1 and S_2 overlap as they share edges (a, b), (a, c), (b, c). S_1 and S_3 are disjoint as they do not share any edge. Similarly S_2 and S_3 overlap. All three subgraphs S_1, S_2, and S_3 are isomorphic as they have the same topology. Notice that isomorphism is a transitive relation. Thus, for a given subgraph S of G, the set of all subgraphs of G that are isomorphic to S defines an equivalence class. We represent the subgraphs in each equivalence class with a graph isomorphic to those in that equivalence class and call it a *pattern*. For example, Fig. 5.5c and d shows the patterns that represent the equivalence classes $\{S_1, S_2, S_3, S_4, S_5, S_6, S_7, S_8\}$ and $\{S_9, S_{10}, S_{11}, S_{12}\}$, respectively. In the rest of this chapter, we will use the terms pattern and motif interchangeably. Also each subgraph within an equivalence class will be termed as *instance* of a motif.

There are two different ways for motif frequency formulation: (1) counting disjoint copies of a motif or (2) allowing different copies of the same motif to overlap (i.e., share nodes or edges). Most of the existing literature on motif counting problem follow the second formulation. This formulation, however, has a fundamental drawback arising from the fact that it does not have *downward closure* property. Briefly, this means that the motif frequency does not decrease or increase monotonically with the motif size. We discuss this drawback in detail later in this chapter. Based on the above two formulation types, there exist alternative definitions of the frequency of a motif in a given graph. The classical frequency definition is the number of all possible subgraphs in the graph which is isomorphic to the given motif. This definition, also known as the $F1$ measure, counts all the subgraphs regardless of whether they overlap with each other or not [42]. There are two other frequency definitions, which avoid overlaps between different isomorphic subgraphs. $F2$ measure counts the size of the largest subset of subgraphs in an equivalence class where no two subgraphs share any edge(s). It, however, allows them to share nodes. $F3$ measure is more stringent as it constrains that no two isomorphic subgraphs can share a node. Consider the graph in Fig. 5.5a and the motif in Fig. 5.5c. The frequency of this motif in the graph according to the $F1$ measure is eight as it has eight embeddings $\{S_1, S_2, \ldots, S_8\}$. On the other hand $F2$ is two with $\{S_1, S_3\}$ or $\{S_1, S_4\}$ or other subsets of two subgraphs with non-overlapping edges, and $F3$ is one (any one from S_1 till S_8).

Now we are ready to discuss the *downward closure* property for motifs. This property states that the frequency of a motif should monotonically decrease as this motif grows in size by inserting new nodes or edges to it. Formally, let us consider two motifs P_1 and P_2 and a function $f()$ that operates on a motif and returns a real number. We say that the function $f()$ has downward closure property if and only if $f(P_2) \leq f(P_1)$ for all (P_1, P_2) pairs where P_1 is a subgraph of P_2. Under the light of these definitions, next, we prove that the $F1$ measure is not downward closed. Consider the motif P_1 in Fig. 5.5b. The frequency of P_1 is three in the target graph of Fig. 5.5a. Now consider the motif P_2 that contains P_1 in Fig. 5.5c. Although P_1 is a subgraph of P_2, the frequency of P_2 is eight in the same graph (i.e., more than that of P_1). Next, consider the motif P_3 in Fig. 5.5d. P_3 contains P_2, and its frequency is only one (i.e., less than that of P_2). This example demonstrates that the $F1$ measure not only fails to monotonically decrease, but it also fluctuates (i.e., its value may increase or decrease) as we grow the motif [48, 53]. Failure to satisfy the downward closure property has major implications on the correctness of motif identification. Traditional motif identification algorithms often grow a motif starting from an initial motif of a small number of edges (see Sect. 5.3.1). Should they employ the $F1$ measure, these algorithms cannot have an early stopping criteria as motif size grows. This is because the frequency may increase as we grow motif size even though the current motif frequency is low.

In all traditional methods, after calculating the number of occurrences of a given motif M in a target network G based on one of the three frequency measures defined above, further computations are necessary to determine the statistical significance of the motif or the frequency of the motif under consideration using various statistical significance testing. First, we count number of occurrences of the same motif using the same frequency measure on a large number of similar random networks (usually $\geq$1000 random networks). Thus we have a large set of frequency values. Let us represent the mean and standard deviation of all those frequency values in the set of random networks with $F_{rand}(M)$ and $\sigma_{rand}(M)$, respectively. We compare these frequency values obtained from random ensembles to the frequency we obtained from G (i.e., $F_G(M)$). We evaluate the significance of the given motif M based on the z-score of its abundance in G against the distribution of its abundance in the random ensembles using the following formula

$$z_{score}(M) = \frac{F_G(M) - F_{rand}(M)}{\sigma_{rand}(M)}$$

The random networks are used to establish default values for frequency measure and other metrics which is known as a statistical null hypothesis. Generated random networks are also known as null models [44]. Several algorithms are used to generate random networks. One of the most famous null models used for this purpose is the Barabási–Albert (BA) model [10]. This model captures the connectivity patterns of real networks [17, 25, 39] and it has been frequently used in the literature to simulate real networks. BA model constructs networks that belong to the class called scale-free (SF) networks whose degree distribution follows the power law. It generates SF network using preferential attachment mechanism. This mechanism states that important nodes (with high degree) are more likely to be connected to a new node compared to other nodes. Erdős Rényi (ER) [18] is also a well known model for random network generation. ER model starts with a set of nodes and connects each node pair with a given probability p. More specifically, each edge in the ER model has fixed probability of being absent or present, independently from other edges.

Canonical labeling is defined as a unique label that is invariant on the ordering of the vertices and edges in the graph [7, 19]. If two subgraphs are isomorphic, then they have the same canonical labeling. The inverse is, however, not true. Unlike isomorphism test, comparing the canonical labeling is a trivial task. It is used by several motif finding algorithms as one of the important solutions of reducing the cost of subgraph isomorphism problem, for example, FSG and FPF methods discussed in Sects. 5.3.1 and 5.3.3, respectively.

5.3 Existing Algorithms

In this section, we discuss various methods that solve the motif identification problem. We classify the literature on discovering motifs based on several orthogonal perspectives that change the problem definition and solution significantly. First, we focus on the number of input networks (i.e., one or more network). Second, we focus on labeling (i.e., labeled or unlabeled) of nodes and edges of the input network(s). Third perspective describes different frequency measures of finding motif instances in a single network. Fourth perspective describes whether the underlying network(s) is directed or undirected. Lastly, we focus on the underlying network model (probabilistic, dynamic). In the following subsections, we discuss each of these perspectives with related case studies.

5.3.1 Number of Input Networks

Several studies in the literature focus on finding motif instances in a single network. The focus lies on analyzing specific network structure and important biological functions in the network under consideration. On the other hand, some other studies are interested in finding motifs across multiple networks. Those studies mostly cluster several networks and based on their topological similarities. In addition, one possible outcome from solving the motif finding problem in multiple networks is to identify important functions that are common in those networks. The number of input networks changes the motive and definition of motif finding problem significantly (for instance, frequency definition of the motif). We discuss two problem definitions and corresponding case studies that solved it in detail later in this section.

5.3.1.1 Single Input Network

In a given single network the problem definition is to find *significant motif instances* (i.e., frequent motif instances that appear more than a given number of times). If the frequency threshold is given, the goal is to find the motifs that have number of instances in the underlying network which exceeds the threshold. This problem becomes much simpler when the motif topology is given. In this case, the problem is to find or count all instances of a given motif in the underlying network. Several methods attempt to solve the motif finding problem using single input network [13, 23, 24, 26, 27, 30, 37, 42, 50]. Next, we discuss one of these methods [23] in detail.

Case Study Given an input graph and an integer size, Grochow et al. [23] proposed an algorithm to find all possible motifs of the given size in the graph. It also reports the significance of abundance of each of those motifs based on z-score. First, the algorithm finds all possible motifs of a given size in the underlying network. For this purpose, it uses existing tools for subgraph enumerations; geng and directg tools [33]. Second, the algorithm then counts number of instances of each of those motifs independently. Instead of enumerating all subgraphs having size as the given input size, it tries to map the motif to the underlying network in all possible ways.

Grochow et al. enhanced subgraph isomorphism test by detecting the degrees of nodes and their neighbors in a subgraph. It starts with two nodes g and h that belong to the original graph and the current motif, respectively, in such a way that g supports h (i.e., there is a possible mapping from h to g based on the degrees of g and h and their corresponding neighbors). It then extends the mapping of those two nodes to include all nodes in the motif topology mapped to some nodes in the original graph by backtracking search. The algorithm uses the most constrained neighbor (to extend mapping with) to eliminate maps that cannot be isomorphic to current motif (i.e., the neighbor of the already-mapped nodes which is likely to have the fewest possible nodes it can be mapped to). Therefore, it selects the nodes having the most already-mapped neighbors. Among those it selects the nodes with the highest degree and largest neighbor degree sequence. When the algorithm successfully maps all nodes from the current motif, the mapped nodes from original graph form a subgraph isomorphic to the current motif. Otherwise, it aborts the mapping as early as possible. The algorithm tries all such combinations of h and g to get all subgraphs that are isomorphic to the current motif.

Considering all combinations of h and g may lead to mapping the same subgraph multiple times to the subgraph symmetry. The authors introduced a new method to avoid exhaustive enumeration of the same subgraph within the underlying network multiple times. This method is called *symmetry-breaking conditions*. It eliminates multiple isomorphism tests of the same subgraph that may happen due to the symmetry in the motif topology. The algorithm computes and enforces several symmetry-breaking conditions, which ensures that there is a unique map from the current motif to each of its isomorphic subgraphs in the original graph, so that exhaustive enumeration search only spends time finding each isomorphic subgraph once. Briefly, the symmetry-breaking conditions are based on node labelings of the current motif by integers, represented as maps from the current motif to the node labelings of graph by distinct integers. Thus, each mapping of the current motif to graph generates a labeling for the nodes in motif. Thus, labelings of motif translate into restraints on maps from motif into graph.

5.3.1.2 Multiple Input Networks

Given a dataset of networks, the problem definition is to find motifs (i.e., subgraphs) such that each appears at least once in a large number of networks in the given dataset. If a motif topology exists in a network among all others in the dataset, its frequency increases by only one for each such network regardless of the number of copies in that network. If a frequency threshold σ is provided as input, then motifs or subgraph topologies should appear in at least σ networks from the input dataset. Several algorithms tackle this problem of finding frequent patterns in multiple graphs. These methods check if the given pattern appears at least once in each graph. Then they report number of networks the motif topology appears in. FSG [29] is one such key method. Vanetik et al. [48] also addressed the same problem. Next, we discuss FSG [29] method to find motifs in multiple networks in detail.

Case Study Given a set of multiple undirected graphs D and a number $\sigma\,(0 \leq \sigma \leq 1)$ as input, FSG [29] solves the problem of finding motifs having frequency cut-off σ. σ represents the ratio of the number of networks that motif appears in to the total number of given networks. FSG finds all undirected subgraphs that occur in at least $\sigma|D|$ of the input graph.

First, the algorithm finds all motifs with one or two edges. Then, based on those two sets, it starts the main computational loop. During each iteration it first generates candidate subgraphs with size greater by one edge than the previous frequent motifs. The method generates all candidate subgraphs of size $(k + 1)$ edges by joining two size (k) edges subgraphs on the condition that those two subgraphs can be

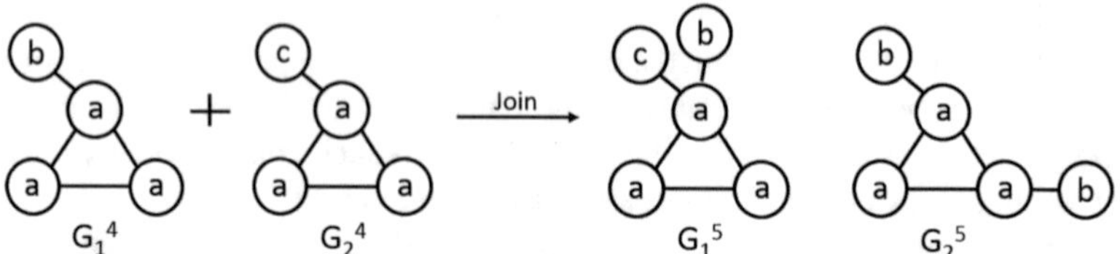

Fig. 5.6 Two subgraphs of size 4 edges joined to form two possible subgraphs of size 5 edges

joined only if they share at least a subgraph of size $(k - 1)$ known as the core subgraph. Figure 5.6 shows potential candidates formed by two structure patterns. Each candidate has one more edge than these two patterns. For each subgraph S with size k edges, the algorithm generates all possible cores (i.e., subgraphs of S with size $(k - 1)$) by removing the edges and check if they exist in previous frequent motifs of size $k - 1$ edges. Next, it counts the frequency for each of these candidates, and prunes subgraphs that do not satisfy the frequency cut-off. Notice that FSG used $F2$ frequency measure. Therefore, removing non-frequent subgraphs in an iteration is safe since these subgraphs will never grow to yield frequent ones in all future iterations. To avoid the costly task of subgraph isomorphism when counting the frequency of a subgraph of size $(k + 1)$, say G^{k+1}, FSG uses Transaction identifier (TID) lists [55]. In this approach it maintains a list of graph identifiers for each frequent subgraph that supports it. It first computes the intersection of the TID lists of its frequent k-subgraphs. If the size of the intersection is below the support, G^{k+1} is pruned. Otherwise it computes the frequency of G^{k+1} using subgraph isomorphism by limiting the search only to the set of graphs in the intersection of the TID lists of its frequent k-subgraphs. Notice that since FSG uses $F2$ measure, due to downward closure property the frequency of a motif decreases if its size increases. Thus, if subgraphs of G^k of size k are not frequent, then G^{k+1} is not frequent either.

FSG uses several algorithms for canonical labeling to uniquely identify the generated subgraphs. This eliminates the computationally expensive subgraph isomorphism tests. One such canonical labeling is based on the labels and degrees of the nodes (i.e., number of incident edges) in the original graph(s). In this approach, the nodes are partitioned into disjointed groups such that each partition contains nodes with the same label and the same degree, and those partitions are sorted by the node degree and label in each partition.

5.3.2 *Graph Labeling*

Several applications require labeling the graph and the motif to only find subgraphs that are isomorphic to a specific motif and have the same labels as well. This is important for finding motif instances to capture the biological process behind this motif as well as its topology. However, most of the existing methods which solve motif finding problem handle the biological networks as uni-labeled graph and only focus on the topological structure of the motif and the network (e.g., [23]). There are many variations of graph labeling perspective. Chen et al. [14] developed one such algorithm. This algorithm considers labelings of the input network and the motif instances. Next, we discuss the LaMoFinder method by Chen et al. that considers labeling of the underlying network vertices.

Case Study Zhou et al. [56] define a Gene Ontology (GO) term as an informative Functional Class (FC) if the GO term has at least 30 proteins directly annotated with it. Following from this, LaMoFinder defines a set $T = \{t_1, t_2, \ldots, t_n\}$ to be the set of GO terms which are assigned to the vertices of network motifs as labels. It states that each GO term in T is either a border informative FC or a descendant of a border informative FC. For each motif g, it denotes the set of occurrences of motif instances in the underlying network G, as D_g. A labeling scheme L of a motif g is said to conform to an occurrence o ($o \in D_g$) if the assigned labels for all vertices of g are either the same or more general than the label of the corresponding vertices in o. LaMoFinder finds all possible labeling schemes for the vertices of g such that these conform to at least σ occurrences in D_g.

LaMoFinder solution extends NeMoFinder algorithm [13] which identifies motif as topological patterns only. LaMoFinder identifies the set of all unlabeled occurrences of g using NeMoFinder algorithm. LaMoFinder then groups those occurrences based on their similarity. It uses informative FC of GO terms as similarity measurements. For each group, it identifies the least general labeling scheme that conforms to all the occurrences in that group. In other words, LaMoFinder tends to select the lowest GO terms that are able to cover all the occurrences.

LaMoFinder uses agglomerative hierarchical clustering method to group the occurrences based on similarity measures. Briefly, this hierarchical clustering works iteratively as follows. In the first iteration, each occurrence forms a cluster. At each iteration, pairs of the most similar clusters are joined to form a new cluster. The algorithm then derives the most specific labeling of the new cluster. The clustering process stops when the labeling scheme has assigned more than half of the vertices with labels that belong to the border informative FC. If the number of occurrences within the cluster exceeds the frequency cut-off σ, then the cluster's labels are saved as a labeling scheme L.

LaMoFinder discovered functionally meaningful motifs of the underlying network. For example, the authors found that a protein with function "carbohydrate utilization" can be regulated by its indirect neighbor with function "regulation of carbohydrate utilization." In addition, they discovered motifs that can be used to predict the functions of unknown proteins in a PPI network. More specifically, LaMoFinder helps to predict unknown functions of a protein represented by a vertex in a motif occurrence g_1 using the corresponding vertex (with known label/function) in the labeled motif occurrence g_2.

5.3.3 Frequency Formulation

In this section, we classify the literature based on frequency measure of a motif in a given network. This is because the frequency measure dramatically changes the cost of counting motifs as well as interpretation of frequency of the underlying pattern. Most of the existing studies use $F1$ frequency measure to count the embeddings of a motif in a given graph (e.g., [13, 23, 26, 27, 37, 50]). These methods have drawbacks that are inherent in the $F1$ measure. First, $F1$ cannot capture the dependency between different instances of the same motif due to the nodes and the edges they share. To understand this better, consider a graph and pattern in Fig. 5.5a and c, respectively. $F1$ counts eight copies of the pattern ($S_1, S_2, \ldots, S_8$). Different nodes and edges, however, contribute multiple times to this count at different embeddings. For instance, the edge (a, b) appears in both S_1 and S_2.

Second and more importantly, the $F1$ measure is not downward closed. This is because as we grow a pattern by including new edges or nodes, its count as computed by $F1$ is not monotonic; it may decrease, stay the same, or increase. Lack of downward closure property makes it impossible to decide if the motif found is the largest one in size while growing a pattern. Thus, using $F2$ or $F3$ is essential for the tractability of identifying frequent patterns.

Finding frequent patterns or counting them without overlaps (i.e., using $F2$ or $F3$ measures) has received little attention in the literature. One of the existing algorithms in this category is SUBDUE [24]. Flexible Pattern Finder Algorithm (FPF) [42] detects frequent patterns using both $F2$ and $F3$. Two algorithms were proposed by Kuramochi and Karypis [30], named hSiGraM, vSiGraM. Next, we select one of these methods, FPF [42], to describe in detail the problem of finding motifs according to $F2$ or $F3$ frequency measures.

Case Study FPF searches for all motifs with a given size and a motif frequency measure ($F1$, $F2$, or $F3$) in a given graph. Following, we discuss their solution for $F2$ and $F3$ only. FPF uses a method that builds a tree of the patterns supported by the target graph and traverses this tree such that only promising branches are examined. The algorithm starts building the tree root that contains the simplest possible pattern with one edge. The tree constructs children by extending the parent pattern by one edge. The algorithm keeps only the frequent patterns and removes all other patterns.

FPF algorithm uses canonical labeling to identify each pattern and collect its isomorphic graphs. Canonical labeling assigns unique labels to the nodes of a given pattern [7]. If two patterns are isomorphic, then they have the same canonical labeling. The inverse is, however, not true. Unlike isomorphism test, comparison of canonical labeling is a trivial task.

The algorithm extracts the Maximum Independent Set (MIS) of each pattern after the determination of all isomorphic subgraphs of this pattern. Since computing MIS is an NP-complete problem [22], FPF uses a heuristic algorithm to compute the $F2$ frequency for a given pattern. This algorithm constructs a new graph, called the *overlap graph* for each pattern as follows. Each node in the overlap graph of a pattern denotes an embedding of that pattern in the target graph.

We add an edge between two nodes of the overlap graph if the corresponding embeddings represented by those nodes overlap in the original graph. Once the overlap graph is constructed, the algorithm starts by selecting the node with the minimum degree (i.e., overlaps with minimum number of embeddings) in the overlap graph. We include the subgraph represented by this node in the edge-disjoint set. We then remove that node along with all of its neighboring nodes from the overlap graph. We update the degree of the neighbors of the deleted nodes. We repeat this process of picking the smallest degree node and shrinking the overlap graph until the overlap graph is empty. To illustrate this, consider the graph G in Fig. 5.5a and the pattern in Fig. 5.5c. This pattern has eight embeddings in G which are $S_1, S_2, \ldots, S_8$. Figure 5.7 represents a partial overlap graph consisting of the subgraphs S_1, S_2, S_3, S_4.

5.3.4 Directed and Undirected Graphs

Recall from Sect. 5.2 that a graph that represents a biological network can be directed or undirected depending on the processes or functions served by this biological network [11]. If the underlying input graph for the motif finding problem changes from directed to undirected graph, the performance and output of the algorithm will change significantly. This is because the number of patterns of the target motif size changes significantly from directed and undirected perspective. We illustrate this by the following example. For instance, the number of possible undirected patterns we can construct with three nodes is two (see Fig. 5.8), while the number of possible directed patterns we can construct with three nodes is 13 (see Fig. 5.9).

Fig. 5.7 A partial overlap graph consisting of subgraphs S_1, S_2, S_3, S_4 of the pattern in Fig. 5.5c

Fig. 5.8 All two possible undirected patterns we can construct with three nodes

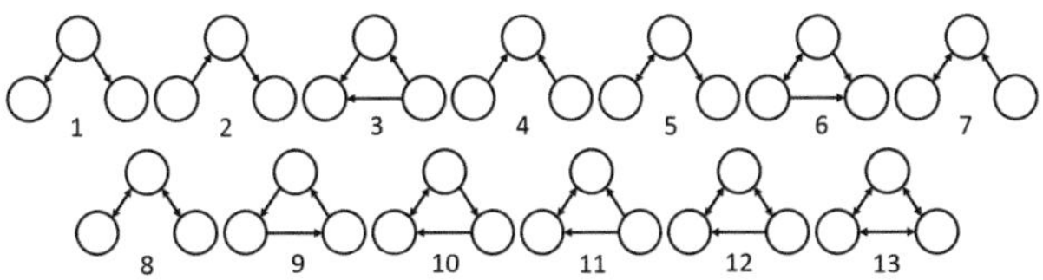

Fig. 5.9 All 13 possible directed patterns we can construct with three nodes

Although having a directed graph introduces extra cost to motif finding problem, most of the existing algorithms can work for both directed and undirected graphs. In other words, the criteria of directed or undirected edges will not change the underlying algorithm significantly. Several algorithms work with directed and undirected graphs (e.g., [23, 26, 27, 37, 40, 42, 50]). Several other algorithms accept only undirected graph as input to their algorithm (e.g., [1, 3, 13]). Next, we select one of these algorithms proposed by Noga et al. [3], to briefly describe an algorithm that find motifs in undirected input graph.

Case Study Noga et al. method counts the number of embeddings of a given motif T with k nodes in a target graph G. The proposed method used *color coding* approach to find instances of the given motif. Color coding approach is based on randomly assigning colors to the nodes of an input graph from a set of k colors. It then considers only those subgraphs where each node has a unique color to detect all subgraphs that may be isomorphic to the given motif. The algorithm then applies dynamic programming to those colorful subgraphs to detect which of those subgraphs are isomorphic to the given motif. This process is repeated many times, $O(e^k)$, to get an estimate of the number of occurrences of the given motif. The algorithm is proved to provide results faster than the algorithm introduced by Grochow et al. [23]. However, this algorithm is limited to motifs of size 10 nodes.

5.3.5 Network Model

We classify algorithms that solve motif finding problem based on network models into three categories: (1) *deterministic model* (algorithms which accept only networks with deterministic topology), (2) *probabilistic model* (algorithms which accept probabilistic networks, i.e., consider a probability value for each of the network edges), and (3) *dynamic model* (algorithms which accept dynamic networks, i.e., different snapshots of a biological network which changes along a specific process). We refer to all the algorithms described in previous subsections in the deterministic model category. Next, we discuss one method each for probabilistic and dynamic model categories in detail.

5.3.5.1 Probabilistic Model

Recall from Sect. 5.1 that interactions within a biological network are not certain all the time. The uncertainty of these interaction is usually represented in the literature as probability values for each edge in the corresponding graph [45]. Probabilistic models to study biological networks help in capturing the uncertainty of the biological processes. More specifically, different physical characteristics exist for multiple deterministic networks which may represent a given probabilistic network. For example, different cancer types exhibit distinctive features (i.e., cell death and growth activities) which signature the genes [21]. The frequency measures described in Sect. 5.3.3 have been used for deterministic networks in the literature. However, this is not the case for probabilistic networks to the best of our knowledge. The reason is that the existence of each embedding depends on the existence of the probabilistic edges which contribute to this embedding. Thus, counting each embedding as one towards its frequency is incorrect due to uncertainty of its edges. Instead, the frequency is interpreted as the expected value of the number of embeddings of the given motif in the underlying probabilistic network.

Few methods in the literature count instances of a given motif in a probabilistic biological network. Hieu et al. [47] introduced a method to solve this problem. This method proposed unique formulations to estimate the count of each motif topology. However, it assumes that all edges have identical failure probabilities. Thus, it fails to solve the generalized version of the problem where each interaction can have a different probability. It counts motif instances according to $F1$ measure but not $F2$. Andrei et al. [46] developed another method which incorporates different interaction probabilities. This method is limited to the $F1$ measure. Next, we briefly discuss one of the existing methods which considers probabilistic interactions to count independent motifs under $F2$ measure.

Case Study Sarkar et al. [41] proposed a novel algorithm to solve motif finding problem in a given probabilistic network according to $F2$ measure (i.e., each motif copy is non-overlapping). Their algorithm starts by hypothetically considering that the underlying network is deterministic. It then counts subgraph topologies according to $F1$ measure. The algorithm then tries to consider the probabilistic interactions while extracting the MIS of each subgraph topology to count motifs according to $F2$ measure.

Unlike the heuristic solution for extracting MIS in a deterministic network (see Sect. 5.3.3), the process of choosing the best node in the overlap graph of a motif does not only depend on its number of neighbors. There are several factors affecting this process. First, the existence of an embedding is not certain since the edges which contribute to this embedding are probabilistic. Thus, we should take into account the probability that the corresponding node in the overlap graph exists. Second and more importantly, the number of neighbors (i.e., number of overlapped embeddings) is not certain as well. Recall from Sect. 5.3.3 that this number is considered as a loss value when choosing the best node to include in the output (i.e., we choose the node with the least loss). Moreover, the probability that each of the neighboring nodes exists is not necessarily independent of each other. This is because neighbors themselves could overlap with each other (i.e., the embeddings denoted by two of the neighbors could have a common edge). Thus, the existence of one depends on another through nontrivial dependencies imposed by the common edges.

The algorithm proposed by Sarkar et al. [41] considers all these dependencies between motif embeddings while counting non-overlapping embeddings (i.e., copies) of each motif topology. Briefly, their algorithm calculates a priority value for each embedding of the given motif. Consider an embedding H_i of a given motif M in the target probabilistic graph G. Their algorithm first calculates a gain value, a_i, of H_i which represents the probability that the corresponding embedding H_i exist. Let us denote the probability of an edge e_j to take place with $P(e_j)$. In their algorithm, they calculate the gain value a_i of an embedding H_i as $P(H_i) = \prod_{e \in H_i} P(e)$. It then calculates a loss value which represents the expected number of overlapped neighbors of H_i given that H_i exists. The priority value of H_i is a function of these two values. After assigning a priority value to each embedding, the algorithm iteratively picks the one with the highest priority. It includes the corresponding embedding to the result set and removes this node and all its neighboring nodes. It repeats this process until the graph is empty.

The complexity of the algorithm lies in calculating the loss value of an embedding H_i of the underlying motif. The challenge of calculating this value is to capture the dependency between the neighbors of each embedding H_k. The authors define a random variable B_i and calculate the distribution of B_i. They compute the loss value of H_i as a function of B_i denoted with $f(B_i)= Exp(Bi)$. The authors calculate B_i in two steps. First, they introduce a new data structure for an embedding H_i to capture the dependency between its neighbors in the overlap graph through the set of edges contributing in those neighbors. This data structure is a bipartite graph $G_i = (\mathcal{V}_1, \mathcal{V}_2, \mathcal{E})$, where $\mathcal{V}_1$ and $\mathcal{V}_2$ represent two node sets, and $\mathcal{E}$ represents the edges connecting nodes of $\mathcal{V}_1$ with those of $\mathcal{V}_2$. Each neighboring node of H_i in the overlap graph corresponds to a node in $\mathcal{V}_1$. Each node in $\mathcal{V}_2$ corresponds to an edge in the edge set, which is the union of edges of all those overlapping embeddings of H_i except for those of the edges of H_i. An edge exists between nodes $u \in \mathcal{V}_1$ and $v \in \mathcal{V}_2$ if the corresponding embedding of node u has the edge denoted by v. To illustrate this, consider the probabilistic graph G in Fig. 5.1a and the pattern in Fig. 5.5b. This pattern has five embeddings in G which are H_1, H_2, H_3, H_4, and H_5 defined by the set of edges $\{(a, b), (a, c), (b, c)\}$, $\{(e, b), (e, c), (b, c)\}$, $\{(d, b), (d, c), (b, c)\}$, $\{(e, d), (e, c), (d, c)\}$, and $\{(e, b), (e, d), (d, b)\}$, respectively. Figure 5.10a represents the overlap graph of this pattern. Figure 5.10b represents bipartite graph of the embedding H_3 of the pattern in Fig. 5.5b.

In the second step, they define a special class of polynomials, called the *x-polynomial* to express this dependency mathematically. They show that the coefficients of this polynomial precisely return the probability distribution of the number of overlapping motif instances with the given instance. The key characteristics of the x-polynomial is that its terms model all possible deterministic network topologies for the edges denoted by $\mathcal{V}_2$. They build the x-polynomial as follows. For each node $v_i \in \mathcal{V}_1$, we define a unique variable x_i. For each node $v_j \in \mathcal{V}_2$, let us denote the existence probability of the edge corresponding to v_j with p_j. Also, let us denote the probability of its absence with $q_j = 1 - p_j$. Using this notation, for each node $v_j \in \mathcal{V}_2$, we construct a polynomial, called the *edge polynomial* Z_j as follows

$$Z_j = p_j \prod_{(v_i, v_j) \in \mathcal{E}} x_i + q_j \tag{5.1}$$

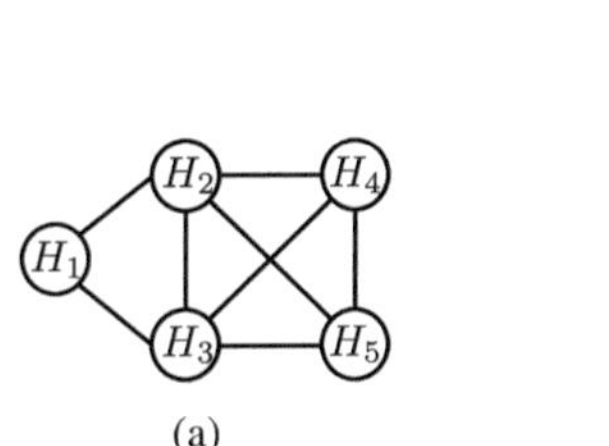

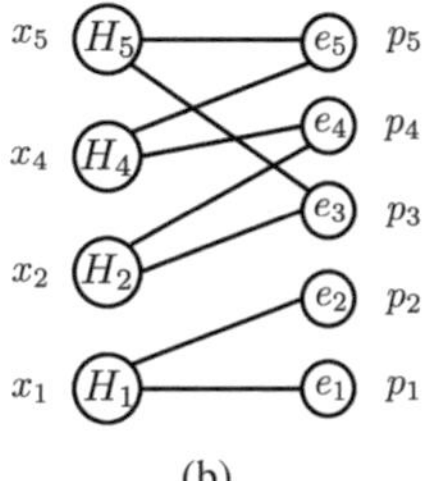

Fig. 5.10 (**a**) The overlap graph of the pattern in Fig. 5.5b in underlying probabilistic network in Fig. 5.1a. (**b**) bipartite graph of the embedding H_3 in the overlap graph in (**a**). $H1$, H_2, H_4, and H_5 are the overlapping embeddings with H_3. Collectively these embeddings yield five edges, which are $\{e_1 = (a, b), e_2 = (a, c), e_3 = (b, e), e_4 = (c, e), e_5 = (d, e)\}$ when excluding edges of H_2. The x-label of each embedding correspond to a unique variable of this embedding. p_i label of an edge e_i represents the probability of the corresponding interaction to take place (i.e., $p_1 = 0.15$ in Fig. 5.1a). Also, let us denote the probability of its absence with $q_j = 1 - p_j$

The first term of this edge polynomial consists of the product of the variables of those overlapping embeddings containing this edge. The second term only has the probability of the absence of this edge. It computes the x-polynomial of H_i as the product of the edge polynomials of all the nodes in $\mathcal{V}_2$. They write the jth term of the x-polynomial as $\alpha_j \prod_{v_i \in \mathcal{V}_1} x_i^{c_{ij}}$, where α_j is the probability and c_{ij} is the exponent of the variable x_i. To illustrate this better, consider the bipartite graph in Fig. 5.10b. The x-polynomial results from this bipartite graph are $(p_1x_1+q_1)(p_2x_1+q_1)(p_3x_2x_5+q_3)(p_4x_2x_4+q_4)(p_5x_4x_5+q_5)$. From this x-polynomial, the authors calculate the expected number of overlapped embedding of H_3, B_3 and consider this value as the loss value when choosing H_3 to put on the output.

The size of this x-polynomial grows exponentially with the number of nodes in set $\mathcal{V}_2$. To reduce the number of terms in this polynomial, their algorithm introduces a collapse operator for each variable x_i denoted with $\phi_r()$, as follows. Let us denote the degree of $v_i \in \mathcal{V}_1$ with $deg(v_i)$. For each node's unique variable x_i, it defines an indicator function $\psi_i(c)$, where $\psi_i(c) = 1$ if $c = deg(v_i)$, otherwise $\psi_i(c) = 0$. Using these notations, for the jth term of the x-polynomial, it defines the collapse as

$$\phi_r(\alpha_j \prod_{v_i \in \mathcal{V}_1} x_i^{c_{ij}}) = \alpha_j[t\psi_r(c_{ij}) + (1 - \psi_r(c_{ij})] \prod_{v_i \in \mathcal{V}_1 - \{v_r\}} x_i^{c_{ij}}$$

Notice that, the collapse operator ϕ_r only changes the variable x_r. It replaces $x_r^{c_{rj}}$ with t if $c_{rj} = deg(v_r)$, which means that all edges of embedding H_r are present (e.g., H_r exists). Otherwise, it completely removes the $x_r^{c_{rj}}$ from this term, which indicates that at least one edge of H_r is absent. It applies the collapse operator ψ_r to the polynomial terms as soon as it completes multiplication of the final edge polynomial of the variable x_r, which means that no other edge polynomials can

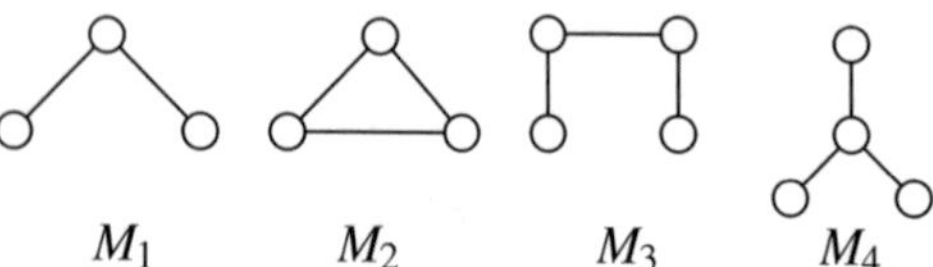

Fig. 5.11 The four motifs with two and three edges

increase the exponent of x_r. After multiplying all edge polynomials and collapsing it, the x-polynomial is in the form:

$$\mathcal{Z}_{H_k} = \sum_{j=0}^{s} p_{kj} t^j \tag{5.2}$$

The coefficients of the polynomial $\mathcal{Z}_{H_k}$ are the true distribution of the random variable B_i (i.e., $\forall k$, the coefficient of t^k is the probability that $B_i = k$).

Once the probability distribution of B_k is calculated, they calculate the expected value of B_i as

$$Exp(B_i) = \sum_{j=0}^{s} j p_{kj} \tag{5.3}$$

Sarkar et al. [41] evaluate their algorithm by finding non-overlapping embeddings of each of all motifs of two or three edges (see Fig. 5.11) on real PPI networks. Those networks already contain interaction confidence values, the authors use them as the interaction probabilities. The results show that their method scales well with growing graph sizes for different network topology models. In addition, they observe that the network topology model has more influence on the performance of their method than the input graph size. This is because the network topology affects the number of overlapping motif embeddings (i.e., neighbors in the overlap graph) for each motif embedding. This in turn increases the polynomial size and affects its processing.

5.3.5.2 Dynamic Model

In this section, we discuss motif finding problem in dynamic networks (see Fig. 5.2). Solving motif finding problem in dynamic network is a very challenging task and received less attention in the literature. One possible way to solve this problem is to consider each snapshot of the underlying network independently and solve the problem from the start. However, this is a bad practice since the change between two snapshots usually does not exceed 5%. In addition, recomputing the frequency from scratch is an expensive task for frequently evolving network.

Case Study The method introduced by Mukherjee et al. [36] is the only method that solves motif finding problem given multiple snapshots of a dynamically evolving biological network. It uses $F2$ measure to count the number of embeddings of each motif. Unlike the trivial solution which computes the frequency of each motif from scratch for each snapshot, their proposed method stores the embeddings of each motif and dynamically process those embeddings when the topology of the network changes. The proposed algorithm considers two possible edit operations which cause change to the current snapshot; edge deletion and edge addition. Recall the algorithm that solves the MIS which uses the concept of the overlap graph. In the first case, the edge deletion, the proposed algorithm considers all embeddings correspond to the nodes of the overlap graph ($F1$ count). The algorithm then checks the list of embeddings from this list that will not exist anymore due to this edge deletion (i.e., all embeddings contain that edge). Let us denote this set with S. The algorithm then collects all embeddings from S which are included in the previous $F2$ count. Let us denote this set with L. Notice that if L is empty, then the $F2$ count will not change and the algorithm stops. If L is not empty, the algorithm first delete all embeddings of L from the previous MIS ($F2$ count decreases by $|L|$). Since all embeddings in L do not exist anymore, the overlapped embeddings with L in S (previous $F1$ count) may be eligible to be included in the non-overlapping set ($F2$ count). The algorithm then includes the embedding from the overlapped embeddings with L that may be included in $F2$ count without violating the non-overlapping condition. If such an embedding is added to the new MIS, the new $F2$ count is increased by one. In the second case, the edge addition, the algorithm collects all new embeddings that are isomorphic to the underlying motif ad results by adding the new edge. Let us denote this set with M. Notice that each embedding in M contains the added edge in the set of its edges. Thus only one embedding from M can be added to the new MIS ($F2$ count). The algorithm finds if any of the embeddings from the set M can be added to the MIS without violating the non-overlapping condition, then the new $F2$ count is increased by one.

5.4 Conclusion

In this chapter, we discussed the problem of finding motifs in biological networks. We briefly introduced the key challenges of solving this problem. We classified the literature of solving this problem based on five different perspectives: the number of input networks, labeling of the nodes and edges, frequency formulation of finding motif instances, directed and undirected input graph(s), and the network model. For each perspective, we briefly summarized one of the exiting methods which handle a variation setting of this perspective. Table 5.1 lists those methods and their variation for each of these perspectives as well as other methods which also solved the problem of finding motifs.

Table 5.1 A survey of the methods we discussed in this work as well as several other methods we mentioned which also solved the problem of finding motifs. Each row represents one of those methods. Each column represents one of the five perspectives of motif identification problem. Each cell has the value of yes if the method considers this perspective or no otherwise or the variation of the method regarding this perspective

	Number of input networks	Graph labeling	Frequency	Edges type	Network model	Motif topology given
Noga et al. [3]	Single	No	$F1$	Undirected	Static	No
NeMoFinder [13]	Single	No	$F1$	Undirected	Static	No
Grochow and Kellis [23]	Single	No	$F1$	Both	Static	Yes
MODA [37]	Single	No	$F1$	Both	Static	Yes
Kavosh [26]	Single	No	$F1$	Both	Static	No
mfinder [27]	Single	No	$F1$	Both	Static	No
FANMOD [50]	Single	No	$F1$	Both	Static	No
Sarkar et al. [41]	Single	No	$F2$	Undirected	Probabilistic	Yes
Mukherjee et al. [36]	Single	No	$F2$	Undirected	Dynamic	Yes
FPF [42]	Single	No	$F1$, $F2$, and $F3$	Directed	Static	No
LaMoFinder [14]	Single	Yes	$F1$	Undirected	Static	No
SUBDUE [24]	Single	Yes	$F2$	Directed	Static	No
FSG [29]	Multiple	Yes	$F2$	Undirected	Static	No

References

1. Ahmed, N.K., Neville, J., Rossi, R.A., Duffield, N.: Efficient graphlet counting for large networks. In: 2015 IEEE International Conference on Data Mining (ICDM), pp. 1–10. IEEE, Piscataway (2015)
2. Albert, I., Albert, R.: Conserved network motifs allow protein–protein interaction prediction. Bioinformatics **20**(18), 3346–3352 (2004)
3. Alon, N., Dao, P., Hajirasouliha, I., Hormozdiari, F., Sahinalp, S.C.: Biomolecular network motif counting and discovery by color coding. Bioinformatics **24**(13), i241–i249 (2008)
4. Ay, F., Dinh, T.N., Thai, M.T., Kahveci, T.: Finding dynamic modules of biological regulatory networks. In: International Conference on BioInformatics and BioEngineering (BIBE), 2010 IEEE , pp. 136–143. IEEE, Piscataway (2010)
5. Ay, F., Kellis, M., Kahveci, T.: SubMAP: aligning metabolic pathways with subnetwork mappings. J. Comput. Biol. **18**(3), 219–235 (2011)
6. Ay, A., Gong, D., Kahveci, T.: Hierarchical decomposition of dynamically evolving regulatory networks. BMC Bioinf. **16**(1), 1 (2015)
7. Babai, L., Luks, E.M.: Canonical labeling of graphs. In: ACM Symposium on Theory of Computing, pp. 171–183 (1983)
8. Bachmaier, C., Brandes, U., Schreiber, F.: Biological networks. In: Handbook of Graph Drawing and Visualization, pp. 621–651. Chapman and Hall/CRC, Boco Raton (2014)
9. Bader, J.S., Chaudhuri, A., Rothberg, J.M., Chant, J.: Gaining confidence in high-throughput protein interaction networks. Nat. Biotechnol. **22**(1), 78–85 (2004)
10. Barabási, A.-L., Albert, R.: Emergence of scaling in random networks. Science **286**(5439), 509–512 (1999)
11. Barabási, A.-L., Oltvai, Z.N.: Network biology: understanding the cell's functional organization. Nat. Rev. Genet. **5**(2), 101–113 (2004)
12. Charlebois, D.A., Balázsi, G., Kærn, M.: Coherent feedforward transcriptional regulatory motifs enhance drug resistance. Phys. Rev. E **89**(5), 052708 (2014)
13. Chen, J., Hsu, W., Lee, M.L., Ng, S.-K.: NeMOfinder: Dissecting genome-wide protein-protein interactions with meso-scale network motifs. In: ACM SIGKDD, pp. 106–115 (2006)
14. Chen, J., Hsu, W., Lee, M.L., Ng, S.-K.: Labeling network motifs in protein interactomes for protein function prediction. In: 2007 IEEE 23rd International Conference on Data Engineering, pp. 546–555. IEEE, Piscataway (2007)
15. Cook, S.A.: The complexity of theorem-proving procedures. In: ACM Symposium on Theory of Computing, pp. 151–158. ACM, New York (1971)
16. Deshpande, M., Kuramochi, M., Wale, N., Karypis, G.: Frequent substructure-based approaches for classifying chemical compounds. IEEE Trans. Knowl. Data Eng. **17**(8), 1036–1050 (2005)
17. Dorogovtsev, S.N., Mendes, J.F.F., Samukhin, A.N.: Structure of growing networks with preferential linking. Phys. Rev. Lett. **85**(21), 4633 (2000)
18. Erdős, P., Rényi, A.: On random graphs. Publ. Math. Debr. **6**, 290–297 (1959)
19. Fortin, S.: The graph isomorphism problem. Technical report, Technical Report 96-20, University of Alberta, Edmonton, Alberta (1996)
20. Gabr, H., Kahveci, T.: Characterization of probabilistic signaling networks through signal propagation. In: Computational Advances in Bio and Medical Sciences, pp. 1–2. IEEE, Piscataway (2014)
21. Gabr, H., Dobra, A., Kahveci, T.: Estimating reachability in dense biological networks. In: ACM Conference on Bioinformatics, Computational Biology and Health Informatics, pp. 86–95 (2015)
22. Garey, M.R., Johnson, D.S.: Computers and intractability: a guide to the theory of NP-completeness. In: Computers and Intractability, p. 340. Macmillan, New York (1979)

23. Grochow, J.A., Kellis, M.: Network motif discovery using subgraph enumeration and symmetry-breaking. In: Research in Computational Molecular Biology, pp. 92–106. Springer, Berlin (2007)
24. Holder, L.B., Cook, D.J., Djoko, S., et al.: Substructure discovery in the subdue system. In: KDD Workshop, pp. 169–180 (1994)
25. Jeong, H., Tombor, B., Albert, R., Oltvai, Z.N., Barabási, A.-L.: The large-scale organization of metabolic networks. Nature **407**(6804), 651–654 (2000)
26. Kashani, Z.R.M., Ahrabian, H., Elahi, E., Nowzari-Dalini, A., Ansari, E.S., Asadi, S., Mohammadi, S., Schreiber, F., Masoudi-Nejad, A.: Kavosh: a new algorithm for finding network motifs. BMC Bioinf. **10**(1), 318 (2009)
27. Kashtan, N., Itzkovitz, S., Milo, R., Alon, U.: Efficient sampling algorithm for estimating subgraph concentrations and detecting network motifs. Bioinformatics **20**(11), 1746–1758 (2004)
28. Krogan, N.J., Cagney, G., Yu, H., Zhong, G., Guo, X., Ignatchenko, A., Li, J., Pu, S., Datta, N., Tikuisis, A.P., et al.: Global landscape of protein complexes in the yeast saccharomyces cerevisiae. Nature **440**(7084), 637–643 (2006)
29. Kuramochi, M., Karypis, G.: An efficient algorithm for discovering frequent subgraphs. IEEE Trans. Knowl. Data Eng. **16**(9), 1038–1051 (2004)
30. Kuramochi, M., Karypis, G.: Finding frequent patterns in a large sparse graph. Data Min. Knowl. Discov. **11**(3), 243–271 (2005)
31. Lee, T.I., Rinaldi, N.J., Robert, F., Odom, D.T., Bar-Joseph, Z., Gerber, G.K., Hannett, N.M., Harbison, C.T., Thompson, C.M., Simon, I., et al.: Transcriptional regulatory networks in saccharomyces cerevisiae. Science **298**(5594), 799–804 (2002)
32. Masoudi-Nejad, A., Schreiber, F., Kashani, Z.R.M.: Building blocks of biological networks: a review on major network motif discovery algorithms. IET Syst. Biol. **6**(5), 164–174 (2012)
33. McKay, B.D.: Isomorph-free exhaustive generation. J. Algorithms **26**(2), 306–324 (1998)
34. Milenković, T., Lai, J., Pržulj, N.: GraphCrunch: a tool for large network analyses. BMC Bioinf. **9**(1), 70 (2008)
35. Milo, R., Shen-Orr, S., Itzkovitz, S., Kashtan, N., Chklovskii, D., Alon, U.: Network motifs: simple building blocks of complex networks. Science **298**(5594), 824–827 (2002)
36. Mukherjee, K., Hasan, M.M., Boucher, C., Kahveci, T.: Counting motifs in dynamic networks. BMC Syst. Biol. **12**(1), 6 (2018)
37. Omidi, S., Schreiber, F., Masoudi-Nejad, A.: MODA: an efficient algorithm for network motif discovery in biological networks. Genes Genet. Syst. **84**(5), 385–395 (2009)
38. Przytycka, T.M.: An important connection between network motifs and parsimony models. In: Research in Computational Molecular Biology, pp. 321–335. Springer, Berlin (2006)
39. Redner, S.: How popular is your paper? An empirical study of the citation distribution. Eur. Phys. J. B Condens. Matter Complex Syst. **4**(2), 131–134 (1998)
40. Ribeiro, P., Silva, F.: G-tries: an efficient data structure for discovering network motifs. In: Proceedings of the 2010 ACM Symposium on Applied Computing, pp. 1559–1566. ACM, New York (2010)
41. Sarkar, A., Ren, Y., Elhesha, R., Kahveci, T.: Counting independent motifs in probabilistic networks. In: Proceedings of the 7th ACM Conference on Bioinformatics, Computational Biology and Health Informatics. ACM, New York (2016)
42. Schreiber, F., Schwöbbermeyer, H.: Frequency concepts and pattern detection for the analysis of motifs in networks. In: Transactions on Computational Systems Biology III, pp. 89–104. Springer, Berlin (2005)
43. Schübeler, D., Scalzo, D., Kooperberg, C., van Steensel, B., Delrow, J., Groudine, M.: Genome-wide DNA replication profile for drosophila melanogaster: a link between transcription and replication timing. Nat. Genet. **32**(3), 438–442 (2002)
44. Shen-Orr, S.S., Milo, R., Mangan, S., Alon, U.: Network motifs in the transcriptional regulation network of Escherichia coli. Nat. Genet. **31**(1), 64–68 (2002)
45. Todor, A., Dobra, A., Kahveci, T.: Characterizing the topology of probabilistic biological networks. IEEE/ACM Trans. Comput. Biol. Bioinf. **10**(4), 970–983 (2013)

46. Todor, A., Dobra, A., Kahveci, T.: Counting motifs in probabilistic biological networks. In: ACM Conference on Bioinformatics, Computational Biology and Health Informatics, pp. 116–125 (2015)
47. Tran, N.H., Choi, K.P., Zhang, L.: Counting motifs in the human interactome. Nat. Commun. **4**, 1 (2013)
48. Vanetik, N., Gudes, E., Shimony, S.E.: Computing frequent graph patterns from semistructured data. In: ICDM, pp. 458–465. IEEE, Piscataway (2002)
49. Wang, P., Lü, J., Yu, X.: Identification of important nodes in directed biological networks: A network motif approach. PloS One **9**(8), e106132 (2014)
50. Wernicke, S.: Efficient detection of network motifs. IEEE/ACM Trans. Comput. Biol. Bioinf. (TCBB) **3**(4), 347–359 (2006)
51. Wuchty, S., Stadler, P.F.: Centers of complex networks. J. Theor. Biol. **223**(1), 45–53 (2003)
52. Wuchty, S., Oltvai, Z.N., Barabási, A.-L.: Evolutionary conservation of motif constituents in the yeast protein interaction network. Nat. Genet. **35**(2), 176–179 (2003)
53. Yan, X., Zhou, X., Han, J.: Mining closed relational graphs with connectivity constraints. In: ACM SIGKDD, pp. 324–333 (2005)
54. Yanover, C., Singh, M., Zaslavsky, E.: M are better than one: an ensemble-based motif finder and its application to regulatory element prediction. Bioinformatics **25**(7), 868–874 (2009)
55. Zaki, M.J.: Scalable algorithms for association mining. IEEE Trans. Knowl. Data Eng. **12**(3), 372–390 (2000)
56. Zhou, X., Kao, M.-C.J., Wong, W.H.: Transitive functional annotation by shortest-path analysis of gene expression data. Proc. Natl. Acad. Sci. **99**(20), 12783–12788 (2002)
57. Zhu, X., Gerstein, M., Snyder, M.: Getting connected: analysis and principles of biological networks. Genes Dev. **21**(9), 1010–1024 (2007)

Chapter 6
Module Identification of Biological Networks via Graph Partition

Yijie Wang

Abstract Complex biological systems can be characterized by biological networks. For example, gene co-expression networks encode relationships between genes, where genes with similar expression profiles are connected to each other. A gene regulatory network is a collection of transcription factors that interact with other genes in the cell to govern the gene expression levels of mRNA and proteins. Obviously, systematic analysis of these biological networks could help us to have a better understanding of the functional organization of the corresponding biological systems and further uncover the underlying biological mechanisms.

Identifying functional modules in these biological networks is one of the key analyses in computational biology. In this chapter, we will introduce the topological property of functional modules and their different definitions. In addition, we will talk about how to convert module identification problems into computational problems. Furthermore, we will introduce how to use computational techniques to solve the module identification problems.

Keywords biological networks · graph partition · modularity · spectral algorithm · semidefinite programming

6.1 Biological Networks

Complex biological systems can be described by biological networks. In general, a node of a biological network represents a biological unit, which could be a cell, a protein, or a gene, and an edge indicates the relationship between those biological units. For different purposes and biological systems, different types of networks are generated. For example, there are protein–protein interaction networks, transcript–

Y. Wang (✉)
Indiana University Bloomington, Bloomington, IN, USA
e-mail: yijwang@iu.edu

© Springer Nature Switzerland AG 2021

B.-J. Yoon, X. Qian (eds.), *Recent Advances in Biological Network Analysis*,
https://doi.org/10.1007/978-3-030-57173-3_6

transcript association networks (gene co-expression networks) and DNA–protein interaction networks (Gene regulatory networks).

For all kinds of theses biological networks, a principle, called guilt-by-association, is widely used. Guilt-by-association declares that the nodes (biological units) in the biological networks, which are connected by an edge, are more likely to perform the same function than nodes (biological units) not linked together. Therefore, it is possible to predict the function of an unknown node (biological unit) through the functions of its topological neighborhood, which have been validated either by chemical or biological experiments.

6.1.1 Protein Interaction Networks

Proteins are large biological molecules, which are composed of one or more connected amino acid residues. They carry out diverse functions within living organisms. Biologically, at both cellular and systemic levels, proteins rarely act alone. Through protein–protein interactions, groups of proteins are organized together to facilitate diverse fundamental molecular processes within a cell.

Protein–protein interactions in a cell can be represented by a protein interaction network, where nodes represent proteins and edges represent protein interactions. Thanks to the high throughput technologies, all protein interactions in a cell can be screened in one test. There are many ways to detect the interactions. The most widely used high throughput methods of detecting protein interactions are yeast two-hybrid screening (Y2H) [22] and affinity purification coupled to mass spectrometry (AP-MS) [5]. Y2H was proposed by Fields and Song in 1989 [6]. Pairwise protein interactions with binary weights can be inferred by Y2H. In Y2H, the transcription factor is separated into two fragments, which are called binding domain and activating domain. The binding domain is fused onto a protein of interesting (referred as the bait protein) and the activating domain is fused onto another protein (referred as the prey protein). If the bait protein and prey protein interact with each other, then the activating domain is brought to the transcription start site, which incurs the occurrence of the transcription of reporter genes. If those two proteins do not interact, then there is no occurrence of the transcription of reporter genes. Based on the occurrence of the transcription of reporter genes, the interactions between proteins can be identified. AP-MS consists of two steps. In the AP (affinity purification) step, a protein of interesting, called bait, is caught in a matrix. The bait protein is passed through the matrix, the protein, interacting with the bait protein, is retained due to interaction with the bait. After purification, proteins can be analyzed by MS (mass spectrometry), which is a chemistry technique that helps to determine the amount and type of chemicals in a sample.

There are many public protein interaction networks database available for researchers and scientists. Some database such as BioGrid [19], DIP [16] and IntAct [7] contain protein interaction networks of many different species. Some databases only maintain protein interaction networks of specific species, such as HPRD hard (a database for human protein interaction network) and FlyBase [4] (a database for fruit fly protein interaction network).

6.1.2 Gene Co-expression Networks

Similar to protein interaction networks, gene co-expression networks are also undirected networks, where each node represents a gene and each edge indicates the similarity of the co-expression patterns of a pair of genes with respect to a bunch of samples. Gene co-expression networks are of biological interest because co-expressed genes are controlled by the same transcriptional regulatory mechanism, or the same pathway or protein complex.

The construction of a gene co-expression network follows a two-step approach. In the first step, the measure of the similarity between every pairs of gene co-expression data is calculated. Then in the second step, edges with small similarities are filtered out by setting a threshold value. The input data for formulating a gene co-expression network is stored in a matrix format. For example, in a microarray experiment, we can obtain the gene expression values of m genes and n samples, then the input data is a $m \times n$ matrix, which is called expression matrix. Each row the expression matrix implies the gene expression pattern of that gene. In the first step, we can estimate the similarity of the gene expression patterns between pair-wise genes through computing the similarity between two row vectors. Pearson's correlation coefficient, Mutual Information, Spearman's rank correlation coefficient, and Euclidean distance are the four mostly used co-expression measures [20]. After the calculation of the similarity scores, we obtain a $m \times m$ similarity matrix, each element of which shows how similarly two genes change together under expression level. In the second step, the elements of the $m \times m$ similarity matrix are binarized based on a certain threshold. The binarized matrix is the adjacency matrix of the gene co-expression network. "1" in the adjacency matrix denotes two genes are correlated under the same samples or conditions, and "0" otherwise.

6.1.3 Cell–Cell Similarity Networks

Recent technological advances have facilitated unprecedented opportunities for studying biological systems at single-cell level resolution. For example, single-cell RNA sequencing (scRNA-seq) enables the measurement of transcriptomic information of thousands of individual cells in one experiment. Analyses of such data provide information that was not accessible using bulk sequencing, which can only assess average properties of cell populations. Specifically, in a scRNA-seq experiment, the gene expression in one cell can be measured. Therefore, we are able to use the output of scRNA-seq data to build a cell–cell similarity network. When constructing the cell–cell similarity network, each node of the network represents a cell and the edges between nodes represent gene expression similarity between cells. The similarity of gene expression patterns between cells can be computed by correlation, mutual information and the most popular one is the cosine similarity. Once we compute the pairwise cell–cell similarity, the k nearest neighbor cell–cell similarity can be built. k nearest neighbor networks play an important role for identify cell types that is the key analysis of scRNA-seq data.

6.2 What Is the Definition of a Module?

We first introduce some terminology to preciously define a module in a network. Assuming we have a network $G(V, E)$, where V is the set of $|V| = n$ nodes representing proteins or genes, and $E = \{e_{ij}\}$ is the set of $|E| = m$ edges, which suggest the physical interactions or correlations. The network G can be presented by an adjacency matrix A, element A_{ij} of which equals to 1 if there is an edge between nodes i and j and 0 otherwise. A_{ij} could also be any non-negative edge attribute. The goal of module identification is to detect a group of nodes, which perform similar functions or possess identical properties. Because we only have topology information of the network (nodes and edges), it is critical to characterize the topological property of a biological meaningful module to guide us find them in a network. In the following, we will discuss several widely used definitions of a functional module.

6.2.1 Modularity

One way to describe a module is based on an observation that the number of edges that connects the nodes in a module should be larger than the number of edges within a group of randomly selected nodes of the same size. More preciously, the module in network $G(V, E)$ should be a group of nodes C such that there are many more edges between the nodes in C than a random module of the same size. Mark Newmen [10] puts such characteristic of a module into a definition called ***Modularity***.

Basically, *Modularity* expresses the relationship between the actual connectivity inside a group of nodes C and the expected connectivity in C. The formal mathematical definition of the modularity of a group of nodes C is

$$Q_C = \sum_{ij} \left[A_{ij} - P_{ij}\right] \delta(g_i, C)\delta(g_j, C), \tag{6.1}$$

where g_i denotes the module that node i belongs to and $\delta(s, t) = 1$ if $r = s$ and 0 otherwise. Therefore, $\delta(g_i, C)\delta(g_j, C)$ equals to one if node i and node j are all in module C and equals to zero otherwise. P_{ij} is the expected number of edges between node i and j (we will discuss how to define P_{ij} in the following). Therefore, Q_C in (6.1) actually calculates the difference between the number of edges among nodes in C and the number of expected edges in C. Larger Q_C indicates that the module has higher internal density than expected. More importantly, we can use Q_C to guide module identification.

On the basis of (6.1), which provides the definition of a module, we are able to find all modules in a given network $G(V, E)$. Assuming the nodes in G can be put into k modules $\{C_1, C_2, \ldots, C_k\}$ and those modules do not have any overlaps, meaning $C_i \bigcap C_j = \varnothing, \forall i, j$. Then we can identify the k modules by maximizing the corresponding modularity as follows:

$$\max : \sum_{ij} \sum_{g_i, g_j} \left[A_{ij} - P_{ij}\right] \delta\left(g_i, g_j\right), \tag{6.2}$$

where $\delta\left(g_i, g_j\right) = 1$ when nodes i and j are assigned to the same module and $\delta\left(g_i, g_j\right) = 0$ otherwise.

Following [10], we define a binary $n \times k$ module assignment matrix H, where the ith row indicates the membership of node i and the jth column presents the jth community. Formally,

$$H_{ij} = \begin{cases} 1 \text{ if node i belongs to community j}, \\ 0 \qquad\qquad \text{otherwise}. \end{cases} \tag{6.3}$$

Note that every row sum of H is 1 and the columns of H are orthogonal.

Noticing that the δ function in (6.2) is equivalent to

$$\delta\left(g_i, g_j\right) = \sum_{l=1}^{k} H_{il} H_{jl}. \tag{6.4}$$

Then (6.2) can be written as

$$\begin{aligned} Q =& \sum_{ij} \sum_{g_i, g_j} \left[A_{ij} - P_{ij}\right] \delta\left(g_i, g_j\right) \\ =& \sum_{i,j=1}^{n} \sum_{l=1}^{k} \left[A_{ij} - P_{ij}\right] H_{il} H_{jl} \\ =& \mathrm{Tr}(H^T W H), \end{aligned} \tag{6.5}$$

where $W = A - P$. There is an implicit constraint for the above problem, which is H is a binary assignment matrix with $\sum_j H_{ij} = 1$ and $\sum_l H_{li} H_{lj} = 0, \forall i \neq j$.

Because we assume all k modules do not overlap, then we can infer that $\sum_j H_{ij} = 1$, meaning node i can be only assigned one module. Using H, we can rewrite (6.2) into a quadratic form as follows:

$$\begin{aligned} \max :& \mathrm{Tr}\left[H^T W H\right] \\ \text{s.t.} & \sum_j H_{ij} = 1, \forall i \\ & H_{ij} \in \{0, 1\}. \end{aligned} \tag{6.6}$$

6.2.1.1 Adjacency Matrix of the Null Model *P*

One key element in the definition of modularity is the use of matrix P. P_{ij} denotes the expected number of edges between nodes i and j. Matrix P can be considered as the weighted adjacency matrix of a null model, which has the same number of nodes as G. To build a good null model, P should satisfy the following constraints. First, P is symmetric, which implies $P_{ij} = P_{ji}$. Second, $Q_C = 0$ when all nodes are placed in a single group. Setting all $g_i \in C$, we have

$$\sum_{ij} \left[A_{ij} - P_{ij}\right] = 0 \Rightarrow \sum_{ij} A_{ij} = \sum_{ij} P_{ij} = 2m. \tag{6.7}$$

Physically, the equation means that the expected number of edges in the entire null model equals to the number of actual edges in G. Additionally, we require the degree distribution of the null model is approximately the same to the original network G. Hence, we need

$$\sum_j P_{ij} = d_i, \tag{6.8}$$

where d_i is the degree of node i. One widely used null model, which satisfies the conditions above is

$$P_{ij} = \frac{d_i d_j}{2m}. \tag{6.9}$$

6.2.1.2 The Resolution Limitation

Modularity suffers from the resolution problem, that is, identifiable modules tend to be merged together if the network is large enough. As an extreme example illustrated in], we could build a network that consists of identical cliques (which are fully connected subgraphs) connected by single links. If the number of the cliques is larger than $\sqrt{m}$, where m is the total number of the edges in the network, then combining two or more cliques would lead to larger modularity (6.2).

The resolution limitation problem arises because of the definition of the null model P. Let us revisit the definition of $P_{ij} = \frac{d_i d_j}{2m}$. We find that if we fix the neighbors of nodes i and j but keeping add new nodes and edges to the rest of the network, then P_{ij} would become smaller and smaller even if their neighbors and local topology remain the same. A concrete example is given in the following.

Assuming we identified three modules $\{A, B, C\}$ using (6.2) for a given network G, where there is only one edge connecting module A and module B. Let us first merge modules A and B and get module AB and further compute the modularity Q_{AB} of module AB based on (6.1).

$$Q_{AB} = \sum_{ij} \left[A_{ij} - P_{ij}\right]\left[(\delta(g_i, A) + \delta(g_i, B)) \times (\delta(g_j, A) + \delta(g_j, B))\right], \tag{6.10}$$

where $(\delta(g_i, A) + \delta(g_i, B)) = 1$ when node i is in the merged module $AB = A \bigcup B$ and similarly $(\delta(g_j, A) + \delta(g_j, B)) = 1$ denotes node j is in the merged module AB. We further expand the above equation and we have

$$\begin{aligned}
Q_{AB} &= \sum_{ij} \left[A_{ij} - P_{ij}\right]\left[(\delta(g_i, A) + \delta(g_i, B)) \times (\delta(g_j, A) + \delta(g_j, B))\right] \\
&= \sum_{ij} \left[A_{ij} - P_{ij}\right]\left[\delta(g_i, A)\delta(g_j, A) + \delta(g_i, B)\delta(g_j, B)\right. \\
&\quad \left. + \delta(g_i, A)\delta(g_j, B) + \delta(g_i, B)\delta(g_j, A)\right] \\
&= \sum_{ij} \left[A_{ij} - P_{ij}\right]\delta(g_i, A)\delta(g_j, A) + \sum_{ij} \left[A_{ij} - P_{ij}\right]\delta(g_i, B)\delta(g_j, B) \\
&\quad + \sum_{ij} \left[A_{ij} - P_{ij}\right]\delta(g_i, A)\delta(g_j, B) + \sum_{ij} \left[A_{ij} - P_{ij}\right]\delta(g_i, B)\delta(g_j, A) \\
&= Q_A + Q_B + \sum_{ij} \left[A_{ij} - P_{ij}\right]\delta(g_i, A)\delta(g_j, B) \\
&\quad + \sum_{ij} \left[A_{ij} - P_{ij}\right]\delta(g_i, B)\delta(g_j, A) \\
&= Q_A + Q_B + 2\sum_{ij} \left[A_{ij} - P_{ij}\right]\delta(g_i, A)\delta(g_j, B).
\end{aligned} \tag{6.11}$$

The fourth equality comes from the modularity definition of modules A and B. And the last equality attains because the network G is symmetric. Clearly, the modularity of Q_{AB} is the summation of Q_A and Q_B plus a third term. Let us further investigate Q_{AB}, we find

$$
\begin{aligned}
Q_{AB} &= Q_A + Q_B + 2\sum_{ij}\left[A_{ij} - P_{ij}\right]\delta(g_i, A)\delta(g_j, B) \\
&= Q_A + Q_B + 2\left[\sum_{ij} A_{ij}\delta(g_i, A)\delta(g_j, B) - \sum_{ij} P_{ij}\delta(g_i, A)\delta(g_j, B)\right] \\
&= Q_A + Q_B + 2\left[1 - \frac{1}{2m}\sum_{ij} d_i d_j \delta(g_i, A)\delta(g_j, B)\right]. \qquad (6.12)
\end{aligned}
$$

The last equality is derived because we assume there is only one edge connecting module A and module B ($\sum_{ij} A_{ij}\delta(g_i, A)\delta(g_j, B) = 1$). We detect modules $\{A, B, C\}$ based on (6.2). Therefore, we know $Q_A + Q_B > Q_{AB}$ and $\frac{1}{2m}\sum_{ij} d_i d_j \delta(g_i, A)\delta(g_j, B) > 1$.

Here comes the interesting part, if we keep adding nodes connecting only to the nodes in module C and keep adding edges between nodes outside module A and B. We notice that the module structures of A and B remain the same, but the total number of edges m in the network G keeps increasing. Then, the increasing m would make $\frac{1}{2m}\sum_{ij} d_i d_j \delta(g_i, A)\delta(g_j, B) < 1$ because m is the only variable that is kept increasing and $d_i d_j, \forall i, j \in AB$ remains the same due to no new edges connecting to modules A and B. Consequently, $\frac{1}{2m}\sum_{ij} d_i d_j \delta(g_i, A)\delta(g_j, B) < 1$ makes $Q_{AB} > Q_A + Q_B$, implying that merge modules A and B would increase the modularity.

6.2.1.3 Multiresolution Modularity

To overcome the resolution limitation, researchers [14] propose the multiresolution versions of modularity. The multiresolution modularity introduces a parameter that controls the resolution. The multiresolution modularity of nodes in module C is defined as

$$
Q_C = \sum_{ij}\left[A_{ij} - rP_{ij}\right]\delta(g_i, C)\delta(g_j, C), \qquad (6.13)
$$

where r works like a resolution parameter: high values of r lead to smaller modules because more penalty is given to P_{ij} to reduce the impact of the total number of edges m in the network. And the module identification formulation based on multiresolution modularity can be written as

$$
\begin{aligned}
&\max : \mathrm{Tr}\left[H^T \left[A - rP\right] H\right] \\
&\text{s.t.} \sum_j H_{ij} = 1, \forall i \\
&H_{ij} \in \{0, 1\}.
\end{aligned}
\tag{6.14}
$$

However, Lancichinetti and Fortunato [8] show that multiresolution modularity still suffers from two coexisting problems: the tendency to combine small subgraphs, which dominates when the resolution parameter r is low; the tendency to break large subgraphs into small parts, which dominates when the resolution parameter r is high. It is still an open question about how to modify the modularity definition to resolve the resolution problems.

6.2.2 Conductance

Another way to define a module is based on its separability. Intuitively, a module could be a group of nodes that rarely connect to the nodes outside of the group. Such property can be characterized by conductance. The mathematical definition of conductance of a group of nodes C is

$$
\phi(C) = \frac{|E(C, \bar{C})|}{\min\{\mathrm{vol}(C), \mathrm{vol}(\bar{C})\}} = \frac{\frac{1}{2}\sum_{ij} A_{ij}\delta(g_i, C)\delta(g_j, C)}{\min\{\mathrm{vol}(C), \mathrm{vol}(\bar{C})\}}, \tag{6.15}
$$

where $\mathrm{vol}(C) = \sum_{i \in C} d_i$ and $|E(C, \bar{C})| = \frac{1}{2}\sum_{ij} A_{ij}\delta(g_i, C)\delta(g_j, C)$ denotes the connections between sets C and $\bar{C} = V - C$. Originally, $\phi(C)$ is used to measure how well the network G is connected. The conductance of the graph G is defined as

$$
\phi(G) = \min_{C \in V} \phi_C. \tag{6.16}
$$

Larger $\phi(G)$ indicates that the graph G is “well-knit.” The definition of conductance can also be used to direct module identification. To identify k modules based on conductance is to solve the following problem:

$$
\begin{aligned}
&\min : \sum_i \phi(C_i) \\
&\text{s.t.} \bigcup_{i=1}^{K} \ldots C_i = V \\
&C_i \bigcap C_j = \varnothing, \forall i, j.
\end{aligned}
\tag{6.17}
$$

Solving (6.17) is challenge because the denominator of $\phi(C_i)$ is a minimization problem $\min\left\{\mathrm{vol}(C_i), \mathrm{vol}(\bar{C_i})\right\}$, which cannot be fixed ahead of time. There is another definition that is very similar to conductance but does not suffer from the problem. The definition is call normalized cut.

For a group of nodes C, the normalized cut of C is defined as [17]

$$N_{cut}(C) = \frac{|E(C, \bar{C})|}{\mathrm{vol}(C)} = \frac{\frac{1}{2}\sum_{ij} A_{ij}\delta(g_i, C)\delta(g_j, C)}{\sum_{i\in C} d_i}. \tag{6.18}$$

Comparing (6.18) with (6.23), obviously, only the denominator is different. When dividing G into large enough k parts $\{V_1, V_2, \ldots, V_k\}$, reasonably assuming at each group $\mathrm{vol}(V_i) < \mathrm{vol}(\bar{V_i})$, then these two definitions are equivalent. Therefore, normalized cut can be viewed as a special case of conductance. If we define the module assignment matrix as H and let h_i represent the ith column of H, then H should satisfy the following constraint to represent the module identification result.

$$\mathfrak{F}_k = \{H : H\mathbf{1}_k = \mathbf{1}_n, H_{ij} \in \{0, 1\}\}. \tag{6.19}$$

$\mathbf{1}_k$ and $\mathbf{1}_n$ are all one vectors with dimension k and n, respectively.

Identification of k modules based on normalized cut can be formulated as

$$\begin{aligned}
\sum_i N_{cut}(V_i) &= \sum_i \frac{|E(V_i, \bar{V_i})|}{\mathrm{vol}(V_i)} \\
&= \sum_i \frac{h_i^T(D-A)h_i}{h_i^T D h_i} \\
&= \mathrm{Tr}\left(\begin{bmatrix} \frac{h_1^T(D-A)h_1}{h_1^T D h_1} & 0 & 0 \\ 0 & \frac{h_2^T(D-A)h_2}{h_2^T D h_2} & 0 \\ 0 & 0 & \cdots \end{bmatrix}_{k\times k}\right) \\
&= \mathrm{Tr}\left[(H^T D H)^{-1} H^T (D-A) H\right].
\end{aligned} \tag{6.20}$$

Our goal is to find a H that can minimize the k-way normalized cuts. The problem can be casted into the following optimization formulation:

$$\begin{aligned}
\min :\ & \mathrm{Tr}\left[(H^T D H)^{-1} H^T (D-A) H\right] \\
\text{s.t.}\ & H \in \mathfrak{F}_k.
\end{aligned} \tag{6.21}$$

Furthermore, we use the definition of Laplacian matrix $L = D - A$ and multiply $D^{-1/2}$ on both size and find out that

$$\begin{aligned} \max: \quad & \operatorname{Tr}\left[H^T D^{-1/2} A D^{-1/2} H\right] \\ \text{s.t.} \quad & H \in \mathcal{F}_k. \end{aligned} \tag{6.22}$$

Another definition that can help us identify community structures in G is conductance [1]. Conductance measures how fast a random walk on G converges to the invariant distribution. For a set of nodes C, its conductance is defined as

$$\phi(C) = \frac{|E(C, \bar{C})|}{\min\{\text{vol}(C), \text{vol}(\bar{C})\}}, \tag{6.23}$$

where $\text{vol}(C) = \sum_{i \in C} \deg(i)$ and $|E(C, \bar{C})|$ denotes the connections between sets C and $\bar{C} = V - C$. Finding S with the minimal conductance is called conductance minimization. In this section, we will discuss several algorithms [1, 17, 21], which are closely related to conductance.

6.2.2.1 Normalized Cut

For a set C, the normalized cut [17] is defined as

$$N_{cut}(C) = \frac{|E(C, \bar{C})|}{\text{vol}(C)} = \frac{\frac{1}{2}\sum_{ij} A_{ij}\delta(g_i, C)\delta(g_j, C)}{\sum_{i \in C} d_i}. \tag{6.24}$$

H is the module assignment matrix and h_i presents the ith column of H. H is in the following constraint set:

$$\mathcal{F}_k = \{H : H\mathbf{1}_k = \mathbf{1}_n, H_{ij} \in \{0, 1\}\}. \tag{6.25}$$

$\mathbf{1}_k$ and $\mathbf{1}_n$ are all one vectors with dimension k and n, respectively.

Detecting k normalized cuts can be formulated as

$$\begin{aligned}
\sum_i N_{cut}(V_i) &= \sum_i \frac{|E(V_i, \bar{V_i})|}{\mathrm{vol}(V_i)} \\
&= \sum_i \frac{h_i^T (D-A) h_i}{h_i^T D h_i} \\
&= \mathrm{Tr}\left(\begin{bmatrix} \frac{h_1^T (D-A) h_1}{h_1^T D h_1} & 0 & 0 \\ 0 & \frac{h_2^T (D-A) h_2}{h_2^T D h_2} & 0 \\ 0 & 0 & \cdots \end{bmatrix}_{k \times k} \right) \\
&= \mathrm{Tr}\left[(H^T D H)^{-1} H^T (D-A) H \right].
\end{aligned} \tag{6.26}$$

Our goal is to find a H that can minimize the k-way normalized cuts. The problem can be casted into the following optimization formulation:

$$\begin{aligned}
\min: \quad & \mathrm{Tr}\left[(H^T D H)^{-1} H^T (D-A) H \right] \\
\text{s.t.} \quad & H \in \mathfrak{F}_k.
\end{aligned} \tag{6.27}$$

Furthermore, we find that the above equation can be further simplified to the following equivalent problem:

$$\begin{aligned}
\max: \quad & \mathrm{Tr}\left[(H^T D H)^{-1} H^T (A) H \right] \\
\text{s.t.} \quad & H \in \mathfrak{F}_k.
\end{aligned} \tag{6.28}$$

Defining $S = \mathrm{Diag}(s_1, s_2, \ldots, s_k) = (H^T D H)^{1/2}$, $Y = D^{1/2} H S^{-1}$, and $W = D^{-1/2}(A) D^{-1/2}$ is the normalized Laplacian matrix, the following problem is equivalent to (6.28):

$$\begin{aligned}
\max \quad & \mathrm{Tr}(Y^T W Y) \\
\text{s.t.} \quad & Y^T Y = I_k, \\
& (D^{-1/2} y_j)_i \in \{0, s_j^{-1}\}, \forall i, j, \\
& Y S \mathbf{1}_k = \mathrm{diag}(D^{-1/2}), \\
& S = \mathrm{Diag}(s_1, s_2, \ldots, s_k) \in R_+^n.
\end{aligned} \tag{6.29}$$

The above problem is a NP-hard combinatorial optimization problem.

6.2.2.2 Block Modeling

Besides looking at the internal density of the module and the external separability of a module, we can also depict a module based on the interaction pattern of the nodes in the module. Basically, the nodes in the module are supposed to have similar interaction patterns. One kind of interaction patterns could be that the nodes in the module highly interact with other nodes of the same module. Another kind of interaction patterns could be that the nodes in the module do not interact but connect nodes in another modules. We can identify modules based on interaction patterns by using the definition of block modeling. Block modeling is more general than modularity and conductance. In this section, we first review block modeling by functional role decomposition proposed by Reichardt and White [15], Reichardt [12], and Pinkert et al. [11].

In block modeling, there is a very important concept called image graph G_B. The image graph is used to present the interactions between modules. Assuming we identified k modules $\mathcal{U} = \{\mathcal{U}_1, \ldots, \mathcal{U}_k\}$ and obtained the module mapping $\tau : V \rightarrow \mathcal{U}$, then we use the adjacency matrix B of the image graph G_B to represent the interaction pattern between modules. $B_{rs} = 1$ indicates there are more edges between modules U_r and $\mathcal{U}_s$ than expectation and $B_{rs} = 0$ denotes the edges between modules $\mathcal{U}_r$ and $\mathbf{U}_s$ are smaller than expectation. Block modeling aims to identify the optimal τ and B to maximize

$$E(\tau, B) = \frac{1}{M} \sum_{i \neq j}^{N} (A_{ij} - B_{\tau_i \tau_j})(A_{ij} - P_{ij}), \tag{6.30}$$

where A_{ij} denotes the weight given to the edge between node i and j (in this paper $A_{ij} = 1$ when node i and j have an interaction, and $A_{ij} = 0$ otherwise); $M = \sum_{i \neq j}^{N} A_{ij}$ is used to bound the error function between 0 and 1; $(A_{ij} - P_{ij})$ denotes the error made on the edge between i and j with P_{ij} as the penalty for the mismatch of the corresponding absent edges. Self-links in the original network are not considered with both $A_{ii} = 0$ and $P_{ii} = 0$. Typically, P_{ij} is chosen the same way as we choose the null model for modularity (6.8).

From equation of $E(\tau, B)$, we find that $B_{\tau_i \tau_j} = A_{ij}$, which means image graph preserves an edge (either existing edge or absent edge) in the original network, leads to $E(\tau, B) = 0$. Otherwise, $B_{\tau_i \tau_j} \neq A_{ij}$, which means image graph does not preserve an edge in original network, leads to $E(\tau, B) = P_{ij}$ when miss-matching the absent edge or $E(\tau, B) = A_{ij} - P_{ij}$ when miss-matching the existing edge. By further investigating $E(\tau, B)$, we find that minimizing $E(\tau, B)$ is equivalent to maximize $\frac{1}{M} \sum_{i \neq j}^{N} (A_{ij} - P_{ij}) B_{\tau_i \tau_j}$ which can be rewritten as $\max_{\tau, B} \frac{1}{2M} \sum_{i \neq j}^{N} (A_{ij} - P_{ij})(2B_{\tau_i \tau_j} - 1)$ by using binary trick. Furthermore, we can formulate the objective function as $Q(\tau, B) = \frac{1}{2M} \sum_{r,s} \sum_{i \neq j}^{N} (A_{ij} - P_{ij}) \delta(\tau_i, r) \delta(\tau_j, s)(2B_{rs} - 1)$. For the original nodes assigned to module node u_r and u_s by τ, we have the corresponding term as $\sum_{i \neq j}^{N} (A_{ij} - P_{ij}) \delta(\tau_i, r) \delta(\tau_j, s)(2B_{rs} - 1)$, in which $\delta(\tau_i, r)$ is the indicator function that takes 1 when $\tau_i = r$ and 0 otherwise. It is clear that the optimal solution for B_{rs} with a given τ is to set $B_{rs} = 1$ when its corresponding term $\sum_{i \neq j}^{N}(A_{ij} - P_{ij})\delta(\tau_i, r)\delta(\tau_j, s)$ is larger than 0, and 0 otherwise. Hence, the optimal solutions of τ and B are naturally decomposed. The optimal image graph B can be derived in a straightforward way once we have the optimal module mapping τ, which maximizes the following equivalent objective function:

$$Q^*(\tau) = \frac{1}{2M}\sum_{r,s}^{q}\left|\sum_{i \neq j}^{N}(A_{ij} - P_{ij})\delta(\tau_i, r)\delta(\tau_j, s)\right|. \tag{6.31}$$

The maximization problem (6.31) is a NP-hard problem [9].

If we use the assignment matrix H to the membership of each node, we can rewrite the blocking modeling as the maximization problem as follows:

$$\begin{aligned} \max : \quad & \left\| H^T(A - P)H \right\|_1 \\ \text{s.t.} \quad & H \in \mathfrak{F}_k. \end{aligned} \tag{6.32}$$

$\|\cdot\|_1$ is the ℓ_1 norm.

Comparing the block modeling formulation (6.32) and the modularity formation (6.2), we notice that maximization modularity is a special case of block modeling. In modularity maximization, we explicitly define the image graph as a network where each node only has self-connection without any other edges connecting to other nodes. The modularity maximization problem (6.2) can be written equivalently as

$$\begin{aligned} \max : \quad & \left\| H^T(A - P)H \otimes B \right\|_1 \\ \text{s.t.} \quad & H \in \mathfrak{F}_k \\ & B = I. \end{aligned} \tag{6.33}$$

$\otimes$ stands for element-wise Hadamard product and I is the identity matrix. Apparently, modularity maximization is a special case of block modeling when we have the prior knowledge of the image graph. Consequently, block modeling also suffers from the resolution limitation problem.

6.3 Graph Partition Algorithms for Module Identification

In previous Sect. 6.2, we introduce different definitions of a module in a network and the graph partition formulations that use those definitions for module detection. In this section, we focus on introducing the state-of-the-art algorithms for solving those graph partition problems.

6.3.1 Algorithms for Modularity Maximization

Modularity maximization (6.2) is a NP-hard problem. There are several approximation methods aiming to solve the problem. Here we will introduce two of them.

6.3.1.1 Spectral Method

We first write out the formulation of modularity maximization.

$$\begin{aligned} \max :&\mathrm{Tr}\left[H^T W H\right] \\ \text{s.t.} &\sum_j H_{ij} = 1, \forall i \\ &H_{ij} \in \{0, 1\}. \end{aligned} \tag{6.34}$$

Because W is a symmetric matrix, hence, it can be diagonalized $W = U\Lambda U^T$, where $U = (u_1, u_2, \ldots)$ is the matrix of eigenvectors of W and Λ is a diagonal matrix of eigenvalues $\Lambda_{ii} = \lambda_i$ $(\lambda_1 \geq \lambda_2 \geq \ldots \geq \lambda_n)$. W may not be positive semidefinite matrix, which means λ_i may be negative. To alleviate the influence of the negative eigenvalues, we do the following transformation:

$$\begin{aligned} Q &= \mathrm{Tr}(H^T B H) \\ &= \mathrm{Tr}(H^T U \Lambda U^T H) \\ &= \mathrm{Tr}(H^T U \alpha I U^T H) + \mathrm{Tr}(H^T U(\Lambda - \alpha I)U^T H) \\ &= \alpha n + \mathrm{Tr}(H^T U(\Lambda - \alpha I)U^T H). \end{aligned} \tag{6.35}$$

Here we make use of the fact that $\mathrm{Tr}(H^T H) = n$. We further remove $n - p$ eigenvectors and compute $\alpha = \frac{1}{n-p}\sum_{i=p+1}^{n} \lambda_i$, then the above equation becomes

$$\begin{aligned} Q &= \alpha n + \mathrm{Tr}(H^T U(\Lambda - \alpha I)U^T H) \\ &\approx \alpha n + \mathrm{Tr}(H^T U_p(\Lambda_p - \alpha I_p)U_p^T H) \\ &= \alpha n + \sum_{l=1}^{k}\sum_{j=1}^{p}\left(\sum_i^n \sqrt{\beta_j - \alpha} U_{ij} H_{il}\right)^2 \end{aligned} \tag{6.36}$$

We define n p-dimensional vectors $r_i = \sqrt{\beta_j - \alpha} U_{ij}$ to characterize each node i in G. Then $\sum_{j=1}^{p}\left(\sum_i^n \sqrt{\beta_j - \alpha} U_{ij} H_{il}\right)^2 = \sum_{j=1}^{p}\left(\sum_{i\in G_l}[r_i]_j\right)^2 = |R_l|^2$, where R_l is the sum of all p-dimensional vectors that belongs to module l. Finally, the modularity Q can be approximate by

$$Q \approx \alpha n + \sum_{l=1}^{k} |R_l|^2. \tag{6.37}$$

Therefore the maximization of modularity converts to clustering the nodes in G into groups so as to maximize the magnitudes of the vectors R_l, which is called vector partition problems. The k-mean algorithm can be easily applied to solve the vector partition problem.

6.3.1.2 Louvain Algorithm

The Louvain algorithm is a hierarchical module identification algorithm that recursively merges modules into a single node and executes the modularity maximization on the condensed graphs.

In order to maximize this modularity (6.1), the Louvain Method has two phases that are repeated iteratively. First, each node in the graph is assigned to its own module. Then for node i, the change in modularity is computed for removing i from its own module and moving it into the module of each neighbor j of i. This modularity is easily calculated by two steps: (1) removing i from its original community and (2) inserting i to the module of j. The two equations are quite similar, and the equation for step (2) is

$$\Delta Q = \left[\frac{\Sigma_{in} + 2d_{i,in}}{2m} + \left(\frac{\Sigma_{tot} + di}{2m}\right)^2\right] - \left[\frac{\Sigma_{in}}{2m} - \left(\frac{\Sigma_{tot}}{2m}\right)^2 - \left(\frac{d_i}{2m}\right)^2\right], \tag{6.38}$$

where Σ_{in} is sum of all the edges inside the module i is moving into, Σ_{tot} is the sum of all the edges of the edges in the module i is moving into, d_i is the degree of node i, $d_{i,in}$ is the sum of the edges between i and other nodes in the module that node i is moving into, and m is the sum of all edges in the network. Then, once this value is calculated for all modules the node i is connected to, node i is placed into the module that generated the greatest modularity increase. If no increase is observed, node i remains in its original module. This process is applied to all nodes. Once this local maximum of modularity is achieved, the first phase ends.

In the second phase of the Louvain algorithm, it puts all of the nodes into the same module and builds a new network where nodes are the modules from the previous phase. Any edges between nodes of the same modules are now presented by self-loops on the new module node, and edges between different modules are represented by weighted edges between new modules. Once the new network is created, the second phase ends and the first phase can be re-applied to the new network.

6.3.2 *PageRank-Nibble Algorithm for Conductance Minimization*

If we define a module based on conductance, then the modules in a network can be detected by finding the minimum conductance sets in the network.

6.3.2.1 Unweighted Network

Andersen [1] developed a local algorithm, called PageRank-Nibble, to find a low-conductance set near a specific starting node in the network based on a personalized PageRank vector. The conductance of the set C identified by the algorithm is at most $f(\phi(C))$, where $f(\phi(C))$ is $\Omega\left(\frac{\phi(C)^2}{\log m}\right)$ (m the number of edges in G). Furthermore, the local algorithm can find C in time $O\left(\frac{m\log^4 m}{\phi(C)^3}\right)$. PageRank-Nibble provides us a power weapon to find a low-conductance near a specific node with theoretical guarantee only spending time linear to the number of edges.

One fundamental step of PageRank-Nibble is to approximate the personalized PageRank vector around node i. Following [1], the lazy variation of PageRank vector is defined as

$$\mathrm{pr}(\alpha, s) = \alpha s + (1 - \alpha)\mathrm{pr}(\alpha, s)\mathcal{P}, \tag{6.39}$$

where α is a constant in (0, 1] called the teleportation constant, s is a distribution called preference vector, and $\mathcal{P} = \frac{1}{2}(I - D^{-1}A)$ is the transition probability matrix of the lazy random walk on G. The personalized PageRank vector used in (6.39) requires $s = e_i$, where e_i is an all zeros vector with one on the ith entry, meaning that the preference vector concentrates on the ith node. The pseudo-code of the approximation is displayed in Fig. 6.1. The technical and theoretical details can be found in [1]. The approximation algorithm can guarantee $\max_{i\in V}\frac{r(i)}{\deg(i)} \geq \xi$.

Once we obtain the approximation of personalized PageRank vector p near node i, then we can sort the nodes around i base on $\frac{p(j)}{\deg(j)}$. Assuming $v_1, v_2, \ldots, v_{Np}$ is the ordering of the nodes around i such that $\frac{p(v_i)}{\deg(v_i)} \geq \frac{p(v_{i+1})}{\deg(v_{i+1})}$, we compute the conductance of the set $C_j = \{v_1, v_2, \ldots, v_j\}$, $j \in \{0, 1, \ldots, N_p\}$. The set C^* with the smallest conductance $C^* = arg\min_{C_j}\phi(C_j)$ is the result produced by the PageRank-Nibble algorithm.

```
Algorithm: ApproximatePageRank (i, α, ξ)
    Let p = 0 and r = e_i.
    While max_{i∈V} r(i)/deg(i) ≥ ξ
        Choose a node i with r(i)/deg(i) ≥ ξ.
        Apply push(i, p, r) and update pr, r.
    Return p ≈ pr(α, e_i).
```

Fig. 6.1 The algorithm to approximate the personalized PageRank vector

```
Algorithm: push (p, r)
    Let p' = p and r' = r.
      p'(i) = p(i) + αr(i).
      r'(i) = (1 − α) r(i)/2.
      For node j (A_ij = 1), r'(j) = r(j) + (1 − α) r(i)/(2deg(i)).
    Return p' and r'.
```

Fig. 6.2 The **push** algorithm

6.3.2.2 Weighted Network

Andersen, et al. [1] only provide the algorithm that aims to deal with unweighted network, where $A_{ij} \in \{0, 1\}$. But in practice, there exist situations that we need to find low-conductance sets for weighted graphs. Here we modify the algorithm based on [1] and propose the weighted versions of Algorithms Figs. 6.1 and 6.2.

Note that the original algorithm was designed for unweighted graphs and we extend the algorithm for weighted graphs as shown in **ApproximatePageRank_weight**(v, α, ϵ), which uses **Push**$_u(p, r)$. ϵ controls the distance between the approximated PageRank vector $\hat{p}$ and the true PageRank vector p.

ApproximatePageRank_weight(v, α, ϵ)

1. Let $p = \mathbf{0}$ and $r = e_v$.
2. While $\max_{u,h \in V} \frac{r(u)A_{uh}}{d(u)} \geq \epsilon$:
   (a) Choose any node u where $\frac{r(u)A_{uh}}{d(u)} \geq \epsilon$
   (b) Apply **Push**$_u$ at node u, and update p and r
3. Return p with $\max_{u,h \in V} \frac{r(u)A_{uh}}{d(u)} < \epsilon$

Push$_u(p, r)$

1. Let $p' = p$ and $r' = r$, except for the following changes:
   (a) $p'(u) = p(u) + \alpha r(u)$
   (b) $r'(u) = \frac{(1-\alpha)}{2} r(u)$
   (c) For each v such that $(u, v) \in E$: $r'(v) = r(v) + \frac{(1-\alpha)}{2} \frac{A_{uv}}{d(u)} r(u)$
2. Return (p', r')

After obtaining the approximated PageRank vector for the weighted network, we can use the same sweep algorithm to find the low-conductance sets introduced in the previous section.

6.3.2.3 Markov Cluster Algorithm

Markov Cluster Algorithm (MCL) has the same flavor as the PageRank-Nibble algorithm. Although there is no proof of equivalence, you may find their similarity once you understand both algorithms.

MCL [21] is a graph clustering algorithm based on manipulation of transition probabilities between nodes of the graph. The underlying principle of MCL is still unknown, which has not prevented its successes on many real-world applications. We categorize MCL as an algorithm related to finding low-conductance sets in network because its similarity to Nibble [18]. Nibble is a local clustering algorithm to recover the low-conductance set C around node i, whose running time is almost-linear with respect to $|C|$. Actually, the PageRank-Nibble algorithm is a variation of Nibble.

MCL iteratively implements "Expand" and "Inflation" operations on the transition matrix $\mathcal{P}$ of the underlying Markov chain of random walk on the given network G. In the "Expand" step, we perform $\mathcal{P}_t = \mathcal{P}_{t-1}\mathcal{P}_{t-1}$. "Inflation" operation follows to compute $\mathcal{P}_t(i, j) = \dfrac{\mathcal{P}_t^r(i, j)}{\sum_{i=1}^{n} \mathcal{P}_t^r(i, j)}$. Those two operations iterate until $\mathcal{P}_t$ converge. Each row of $\mathcal{P}_t$ contains the membership information corresponding to one cluster. Generally, most of the rows of $\mathcal{P}_t$ converge to all zero vectors. The only parameter of MCL is r, which control the size of the modules in average sense. If r is large, then the module size tends to become small.

On comparison, Nibble also consists of two major steps, which is random walk propagation and removal of unrelated nodes. Nibble tries to compute the random walk probability vector within the first several steps. It starts with vector $q_0 = e_v$. In the random walk propagation step, $q_t = \mathcal{P}q_{t-1}$ is performed. In the removal step, the nodes with probabilities smaller than $\varepsilon \cdot \deg(i)$, where $\varepsilon = \phi^2/(\log^3 m2^b)$, are zero out.

Comparing MCL with PageRank-Nibble, intuitively, both of them have two similar steps, random walk propagation and node deletion. The similarity may explain why MCL yields reasonable results.

6.3.3 *Algorithms for the Normalized Cut Problem*

Normalized cut problem is one of the well investigated graph partition problem. Here we introduce several classic algorithms to solve it.

6.3.3.1 Spectral Method

A spectral method can be used to approximate the solution of (6.29). Consider the following relaxed problem of (6.29):

$$\begin{aligned} \max \quad & \mathrm{Tr}(Y^T W Y) \\ \text{s.t.} \quad & Y^T Y = I_k. \end{aligned} \tag{6.40}$$

Comparing with the original un-relaxed problem (6.29), we notice that we remove all other constraints except the orthogonal constraint.

Although $\mathrm{Tr}(Y^T W Y)$ is not convex because W may not be positive semidefinite matrix, we can retrieve the optimal solution supported by Kay Fan theorem. Based on the theorem, the optimal Y^* is attained at $Y^* = U_k$, whose columns are the eigenvectors corresponding to the k largest eigenvalues of W. k-means clustering can be used to obtain a feasible solution in the original constraint set.

Ky Fan Theorem *Let T be a symmetric matrix with eigenvalues $\lambda_1 \geq \lambda_2 \geq \ldots \geq \lambda_n$ and the corresponding eigenvectors $U = (u_1, \ldots, u_n)$. Then*

$$\sum_{i=1}^{k} \lambda_i = \max_{X^T X = I_k} \mathrm{Tr}(X^T T X). \tag{6.41}$$

Moreover, the optimal X^* is given by $X^* = [u_1, \ldots, u_k]Q$ with Q being an arbitrary orthogonal matrix.

6.3.3.2 Semidefinite Programming Method

The normalized cut problem can also be relaxed to a semidefinite programming problem (SDP). The solution to the SDP problem can be computed optimally and efficiently and, after additional adjustment, provides an approximate solution to the original problem. Based on previous results Xing:CSD-03-1265, we know the original problem (6.29) can be written as

$$\begin{aligned} \min: \quad & \mathrm{tr}(L_A Z) \\ \text{s.t.} \quad & Z\mathrm{diag}(D^{-1/2}) = \mathrm{diag}(D^{-1/2}) \\ & \mathrm{tr}(Z) = k \\ & Z = Z^T, Z \succeq 0 \\ & Z_{ij} \geq 0, \forall i, j. \end{aligned}$$

$L_A = I - D^{-1/2}AD^{-1/2}$ is the normalized Laplacian matrix. $Z = YY^T$ and $Y = D^{-1}Y(Y^T DY)^{-1/2}$. The above formulation can be solved by SDPe.

In general, due to relaxation, the optimal solution of SDP does not represent a feasible answer to our original problem (Eq. (6.29)). Therefore, we need to recover a closest feasible solution to the original problem (6.29). To this end, we treat row i of Z (the optimal solution of (SDP)) as the topological feature of node i. Then we apply k-means to Z 100 times and pick one k-means solution which yields the minimum objective function value of (6.29).

6.3.4 *Algorithms for Block Modeling*

Comparing to modularity maximization, identification of low-conductance sets and the normalized cut problem, block modeling is computationally more challenging to solve.

6.3.4.1 Simulated Annealing

The first algorithm that tried to solve (6.32) is the simulated annealing method. However, in order to obtain quality result, we need the cooling down procedure to be sufficiently slow. Therefore, for large scale biological network the practical computational time is unacceptable. The details of the simulated annealing method can be found in [13].

6.3.4.2 Subgradient-Based Optimization Algorithm

We note that the optimization problem (6.33) is a non-smooth combinatorial optimization problem as the objective function involves the L_1 norm of the matrix $Q = H^T(A - P)H$.

$$\begin{aligned} \max: \quad & F(H) := \left\| H^T(A-P)H \right\|_1 \\ \text{s.t.} \quad & H \in \mathcal{F}_k. \end{aligned} \tag{6.42}$$

To solve this hard optimization problem, we first relax the binary constraints $H \in \{0, 1\}^{N\times k}$ in (6.33) by continuous relaxation $H \in [0, 1]^{N\times k}$ and define Φ to represent the relaxed constraint set, which is a convex hull after relaxation.

$$\Phi = \{H : H\mathbf{1}_k = \mathbf{1}_n, H_{ij} \in [0, 1]^{N\times k}\}. \tag{6.43}$$

To address the nonlinearity of the matrix L_1 norm objective function $F(H) = -\|Q\|_{L_1}$ with the relaxed linear constraints, we propose to use Frank–Wolfe algorithm [3] to iteratively solve the following optimization problem with a linear objective function from the approximation by the first-order Taylor expansion:

$$\begin{aligned} &\min_S : F(H^t) + < \nabla F(H^t), (H - H^t) > \\ &\text{s.t.} \;\; H \in \Phi, \end{aligned} \tag{6.44}$$

where H^t is the current solution, $< , >$ is the inner product operator, and the new objective function is from the first-order Taylor expansion. The problem (6.44) at each iteration is a linear programming problem to search for the local extreme point along the gradient $\nabla F(H^t)$ as in steepest descent. However, as previously stated, $F(H^t)$ takes the matrix L_1 norm, which is non-smooth, and therefore non-differentiable. To address this last complexity, we apply subgradient methods [3] to replace $\nabla F(H^t)$ by a subgradient $\partial F(H^t)$ instead [2]:

Definition (Subgradient) A matrix $\partial F \in \mathcal{R}^{m\times n}$ is a subgradient of a function $F : \mathcal{R}^{m\times n} \to R$ at the matrix $X \in \mathcal{R}^{m\times n}$ if $F(Z) \geq F(X) + < \partial F, (Z - X) >$, $\forall Z \in \mathcal{R}^{m\times n}$.

In our case, the subgradient of the matrix L_1 norm can be presented by its dual norm—matrix L_∞ norm, which is used to derive the subgradient $\partial F(S^t)$. Similar to the derivation for the subgradient of the L_1 norm of vectors in L_1 regularization in [2], we show that the subgradient of the L_1 norm of any matrix X is

$$\partial \|X\|_{L_1} = \begin{cases} \left\{Y \in \mathbf{R}^{m\times n}; \|Y\|_{L_\infty} \leq 1)\right\}, & \text{if } X = \mathbf{0}; \\ \left\{Y \in \mathbf{R}^{m\times n}; \|Y\|_{L_\infty} \leq 1 \; and \; < Y, X >= \|X\|_{L_1}\right\}, & \text{otherwise}, \end{cases} \tag{6.45}$$

where $\mathbf{0}$ is a $m \times n$ matrix of all zeros. For our module identification problem, we have the following proposition derived from (6.45):

Proposition *The subgradient of the objective function of our relaxed optimization problem $F(S)$ at the assignment H^t can be defined as: $\partial F(H^t) = 2(P - W)H^t\overline{Q}$. In our implementation, we choose*

$$\overline{Q}_{rs} = \begin{cases} \alpha & Q_{rs} = 0; \\ 1 & Q_{rs} > 0; \\ -1 & Q_{rs} < 0, \end{cases} \tag{6.46}$$

where α is a number between $[-1, 1]$.

Proof From (6.45), there always exists a $\overline{Q}$ satisfying $\|\overline{Q}\|_{L_\infty} \leq 1$ and $\|Q\|_{L_1} =< \overline{Q}, Q >$. As $\partial \|Q\|_{L1} = \partial < \overline{Q}, Q >$ and the subgradient of differentiable functions is equal to its gradient [2], we have $\partial F(H^t) = -\partial \left[\|Q\|_{L1}\right] = -\partial < \overline{Q}, Q >= -\partial \text{tr}(\overline{Q}^T {H^t}^T (W - P) H^t) = 2(P - W) H^t \overline{Q}$ when H^t is close to local minima. QED.

Using Frank–Wolfe algorithm with the derived subgradient, we now have a conditional subgradient method [2] to iteratively solve the relaxed optimization problem as shown in the pseudo-code given in the following **Algorithm**.

Algorithm: Conditional subgradient

Input: initial value S^t, $t = 0$.

Do:

(i) Compute the subgradient $\partial F(S^t)$.

(ii) Solve the minimization problem:

$$S^* = \arg\min_S : \quad < \partial F(S^t), S > \quad \text{s.t.} \quad S \in \gamma$$

(iii) Linear search for the step in the direction $S^* - S^t$ found in (ii), update S^t, $t = t + 1$.

Until: $|\Delta F| + \|\Delta S^t\| < \xi$

Output: S^t.

In this algorithm, step (ii) at each iteration can be solved using a generic linear programming solver in $O((qN)^{3.5})$. However, due to the special structure of the optimization problem, we instead solve it as a semi-linear assignment problem(because assignment matrix $[\partial F(H^t)]_{N \times k}$ is not a square matrix), which can be efficiently solved by assigning node i to module r, which is the index of the largest entry in row i of subgradient $\partial F(H^t)$ with the time complexity $O\ (N \times k)$.

In order to make use of the local optima found by the fast subgradient method, the path generation technique can be used to search for better solution of combinatorial optimization problem by using local optima. The path generation method aims to conserve the overlaps between two local optima results, and get improvement based on the overlaps which make great contribution to the objective value. Our path generation is inspired by path relinking method which connects two combinatorial local optima and try to find better result along the connection [9].

The essential idea of the path generation method is to construct new results by preserving useful overlaps between modules from two local optima. Given two solutions x_A and x_B as the new path generators, PG generates new results and explores the search space on basis of maintaining the current productive overlaps between x_A and x_B. Let $N_r(x_A)$ denote the module u_r of x_A and $N_s(x_B)$ the module u_s of x_B. The contribution $H(r, s)$ by maintaining the overlap $Over(r_A, s_B)$ between $N_r(x_A)$ and $N_s(x_B)$ is defined as:

$$H(r, s) = \|h_{AB}^T(W - P)H_A\|_{L1} + \|h_{AB}^T(W - P)H_B\|_{L1} \tag{6.47}$$

in which h_{AB} is a binary vector, of which each element is equal to 1 when the corresponding node is in both $N_r(x_A)$ and $N_s(x_B)$, and 0 otherwise. The value of $H(r, s)$ is the shared contribution to the objective function Q^* in (6.31) between $N_r(x_A)$ and $N_s(x_B)$ in two feasible solutions. H_A and H_B are assignment matrix of the two solutions. Then the most promising overlap between modules r_A and s_B are determined by

$$(r_A, s_B) = argmax\{H(r, s) : r, s \in \{1, \ldots, q\}\}. \tag{6.48}$$

The path generation based on (6.48) proceeds in the following manner: First, the most promising overlap $Over(r_A, s_B)$ between modules r_A and s_B of the initiating solution x_A and the guiding solution x_B are identified by (6.48), then r_A is locally adjusted to become $Over(r_A, s_B)$ by removing nodes. After the adjustment, a new solution x_1 is generated and $C_A = \{r_A\}$ and $C_B = \{s_B\}$, where C_A and C_B denote the sets of used modules in both solutions. Local search is then applied to find the improved x_1^*. Then we preserve x_1^* and let $x_A = x_1^*$ and repeat the above procedure until no overlap exists or other termination condition (for example, $N_{stop} = 5$ means that there are no larger than 5 nodes overlap modules exist in both solutions). In the end, we obtain the best solution along the generated results. The whole procedure is illustrated as follows:

Algorithm: Path generation method

Input: $x_A, x_B, x, x_{best}, N_{stop}, C_A = \emptyset, C_B = \emptyset, Over = +\infty, Q_{best} = -\infty$

While($Over > N_{stop}$)

(1) $(r_A, s_B) = argmax\{S(r, s) : r, s \in \{1, \ldots, q\}\}$ and find $Over(r_A, s_B)$;

(2) modify nodes from r_A in x to make $N_r(x) = Over(r_A, s_B)$ and $C_A = \{r_A\}, C_B = \{s_B\}$;

(3) $(Q_x^*, x^*) =$ LocalSearch(x);

(4) **If** ($Q_x^* > Q_{best}$)

(5) $Q_{best} = Q_x^*$ and $x_{best} = x^*$;

(6) **EndIf**

(7) $x_A = x^*$ and find the next $Over$ set using (6.48);

EndWhile

Output: x_{best} and Q_{best}.

6.4 Conclusion

In this chapter, we reviewed different definitions of a module in a biological network and the corresponding module identification formulations built on these definitions. In addition, we reviewed the classic algorithms that are designed to solve those module identification problems. What we introduced in this chapter is just part of the puzzle. But it is a good starting point if you want to dedicate your research in this area. We believe more effort would be made in this area in the near future because module identification is the stepping stone for analyzing biological networks.

References

1. Andersen, R., Chung, F., Lang, K.: Local graph partitioning using PageRank vectors. In: 2006 47th Annual IEEE Symposium on Foundations of Computer Science (FOCS'06), pp. 475–486 (2006)
2. Bash, F., Jenatton, R., Mairal, J., Obozinski, G.: Optimization with sparsity-inducing penalties. Found. Trends Mach. Learn. **4**(1), 1–106 (2012)
3. Bertsekas, D.P.: Nonlinear Programming, 2nd edn. Athena Scientific, Belmont (1999)
4. Drysdale, R.: FlyBase: A database for the Drosophila research community. Methods Mol. Biol. **420**, 45–59 (2008)
5. Dunham, W.H., Mullin, M., Gingras, A.C.: Affinity-purification coupled to mass spectrometry: basic principles and strategies. Proteomics **12**(10), 1576–1590 (2012)
6. Fields, S., Song, O.: A novel genetic system to detect protein-protein interactions. Nature **340**(6230), 245–246 (1989)
7. Kerrien, S., Aranda, B., Breuza, L., et al.: The intact molecular interaction database in 2012. Nucleic Acids Res. **40**(D1), D841–D846 (2012)
8. Lancichinetti, A., Fortunato, S.: Limits of modularity maximization in community detection. Phys. Rev. **E84**, 066122 (2011)
9. Mateus, G.R., Resende, M.G.C., Silva, R.M.A.: GRASP with path-relinking for the generalized quadratic assignment problem. J. Heuristics **17**, 527–565 (2011)
10. Newman, M.E.J.: Finding community structure in networks using the eigenvectors of matrices. Phys. Rev. E **74**, 036104 (2006)
11. Pinkert, S., Schultz, J., Reichardt, J.: Protein interaction networks: More than mere modules. PLoS Comput. Biol. **6**, e1000659 (2010)
12. Reichardt, J.: Structure in Networks. Springer, Berlin (2008)
13. Reichardt, J.: Structure in Complex Networks. Springer, Berlin (2009)
14. Reichardt, J., Bornholdt, S.: Statistical mechanics of community detection. Phys. Rev. E **74**, 016110 (2006)
15. Reichardt, J., White, D.R.: Role modules for complex networks. Eur. Phys. J. B **60**, 217–224 (2007)
16. Salwinski, L., Miller, C.S., Smith, A.J., Pettit, F.K., Bowie, J.U., Eisenberg, D.: The Database of Interacting Proteins: 2004 update. Nucleic Acids Res. **32**, D449–D451 (2004)
17. Shi, J., Malik, J.: Normalized cuts and image segmentation. IEEE Trans. Pattern Anal. Mach. Intell. **22**(8), 888 (2000)
18. Spielman, D., Teng, S.-H.: Nearly-linear time algorithms for graph partitioning, graph sparsification, and solving linear systems. In: Stoc 2004 (2004)
19. Stark, C., Breitkreutz, B.J., Reguly, T., Boucher, L., Breitkreutz, A., Tyers, M.: BioGRID: A general repository for interaction datasets. Nucleic Acids Res. **34**, D535–D539 (2006)

20. Stuart, J.M., Segal, E., Koller, D., Kim, S.K.: A gene-coexpression network for global discovery of conserved genetic modules. Science **302**(5643), 249–255 (2003)
21. Van Dongen, S.: A cluster algorithm for graphs. Technical Report INS-R0010 (2000)
22. Young, K.: Yeast two-hybrid: so many interactions, (in) so little time. Biol. Reprod. **58**(2), 302–311 (1998)

Chapter 7
Network Module Detection to Decipher Heterogeneity of Cancer Mutations

Yoo-Ah Kim

Abstract The analysis of genomic landscapes of cancer patients taken from large-scale datasets revealed the complexity and heterogeneity of the disease. Namely, the patients even with the same disease subtypes may have very different genomic profiles and different treatment outcomes. For example, the variations in drug response among patients can be caused by differences in mutation patterns, and understanding the relationship is crucial to design personalized medicine. Such heterogeneity in cancer can be better understood from the perspective of pathways by acknowledging the fact that genes altered in different patients may belong to the same pathway and, therefore, lead them to have similar disease subtypes. Their responses to a drug may still vary due to the differences in genetic alteration status and interactions between the genes. Computational methods to understand the mutational patterns harbored by the genes and their interplay in the context of functional interactions can help decipher cancer heterogeneity. In this book chapter, we discuss various utilities of networks in cancer studies, ranging from detecting cancer driver modules, uncovering modules associated with specific cancer phenotypes, to understanding the relationship between gene modules in different pathways.

Keywords Cancer heterogeneity · Mutual exclusivity · Co-occurrence · Functional interaction network · Cancer driver mutations · Genotype–phenotype association · Set cover

7.1 Introduction

The emergence of large-scale cancer databases offered unprecedented opportunities to analyze genomic landscapes of cancer systematically. The databases catalogue

Y.-A. Kim (✉)
National Center of Biotechnology Information, National Library of Medicine, NIH, Bethesda, MD, USA
e-mail: kimy3@ncbi.nlm.nih.gov

© Springer Nature Switzerland AG 2021

B.-J. Yoon, X. Qian (eds.), *Recent Advances in Biological Network Analysis*,
https://doi.org/10.1007/978-3-030-57173-3_7

comprehensive genomic, epigenomic, and transcriptomic data for thousands of cancer samples with the goal to better understand the genetic causes and to advance cancer diagnosis, cure, and treatments. In addition, gene interaction networks can provide the information on functional and physical relationships among genes. Computational tools have been developed to analyze and integrate multiple data types and answer critical questions in cancer such as identifying cancer drivers [4, 21, 29, 33, 36, 37], classifying cancer subtypes [12, 15], and predicting survival time or drug responses [18, 24].

Although the systematic analyses of large-scale cancer genome datasets significantly improved our understanding of genetic basis of cancer causing mechanisms, many questions still remain not fully answered. In particular, cancer genomes manifest heterogeneous characteristics in multiple aspects, making the interpretation challenging. Each cancer genome evolves in its own unique way, and mutational profiles differ greatly among different cancer genomes (i.e., intertumor heterogeneity) even if they belong to the same cancer subtypes. Although not being discussed in detail in this work, cancer heterogeneity has been also observed among different cells in the same genome (i.e., intratumor heterogeneity), which can provide clues to the evolutionary trajectory of a tumor. Below, we summarize several unique properties of cancer mutations that systematic analyses of cancer genomes revealed.

Mutual Exclusivity It is often observed that the sets of cancer patients with mutations in two cancer driving genes are mutually exclusive. One of possible explanations on the property is that the two genes belong to the same cancer driving pathway and only one mutation in the pathway is sufficient to dysregulate the function of the pathway. Building on the idea, mutual exclusivity properties [4, 8, 11, 12, 21, 28–30] have been extensively studied for the identification of cancer drivers. Such mutually exclusive patterns, however, can also arise due to different reasons or among noncancer driving mutations. For example, mutual exclusivity might reflect mutations specific to two different subtypes or occur due to genetic interactions such as synthetic lethality. There have been several algorithms developed to accurately measure the significance of mutual exclusivity or classify the types of mutual exclusivity [21, 23, 31, 34].

Co-occurrence Another property that is often observed in cancer mutation is co-occurrence, which indicates that a set of genes have mutations in the same subset of patients. As with mutual exclusivity, co-occurrence of mutations can occur due to different reasons. For example, copy number variations often occur in the regions affecting multiple genes, causing genes nearby to have co-occurring alterations. Mutational signatures (see more discussion below) may create co-occurring mutations in the genes when they reside in the regions susceptible to specific mutagenic processes. Another possibility is functional redundancy, in which case a mutation in either gene may not have any effect, but mutations in both genes lead to cancer.

Long Tail Distribution of Cancer Driver Mutations While a few cancer driver genes are highly mutated (>20% of samples), the analysis of thousands of cancer genomes found that there may be hundreds of potential cancer drivers with low or medium mutation frequencies, exhibiting a long tail distribution [3, 27]. The property demonstrates that analyzing more cancer samples may be required to discover a more comprehensive list of cancer driver genes. At the same time, functional analysis based on network information may be helpful to distinguish cancer driving mutations with low mutation frequencies from passenger mutations. It is difficult to identify such rare mutations using traditional methods, and network-based approaches come handy in identifying such driver mutations as considering functionally related genes together can provide an additional statistical power.

Mutational Signatures Cancer genomes naturally accumulate mutations over time, most of which are benign passenger mutations. Although they may not cause cancer directly, those patterns can offer important clues to tumor development. Mutations can be caused by either endogenous or exogenous sources. Endogenous mutagenic processes are related to malfunctioning of DNA repair and maintenance mechanisms. Mutations can also be caused due to environmental factors such as excessive UV light exposure, tobacco consumption, etc. Different mutagenic processes tend to leave their characteristic mutational patterns, and recent computational analyses successfully decomposed mutational profiles for a large number of cancer genomes into multiple different patterns, called "mutational signatures." Each cancer genome contains different levels of mutations in each of the mutational signatures. Some of the mutational signatures are linked to specific mutagenic processes. Understanding the etiology of mutational signatures can provide the information on which cancer causing mutagenic processes a cancer genome has undergone, suggest clues to understand the mechanism of tumorigenesis, and help predict effective personalized cancer treatment.

Finding mutations associated with the disease is fundamental to uncover the genetic causes of cancer development and other cancer phenotypes. However, the heterogeneity and other characteristics of cancer mutations described above pose substantial challenges in delivering the tasks and call for sophisticated computational tools. In this work, we discuss the computational methods to identify genetic alterations causing cancer or cancer-related phenotypes and understand their complex relationships, taking into account cancer heterogeneity. In particular, one of the popular approaches is a set cover-based method, which chooses cancer-related genes while acknowledging the fact that different patients can have different mutations. Given the distinct characteristics of cancer mutations, network-based approaches can also be useful in answering various questions to better understand cancer mutations. Cancer heterogeneity may arise due to the fact that cancer is a disease of dysregulated pathways rather than a mutation in a single gene. Therefore, combining network information offers an additional advantage to decipher the heterogeneity by considering functionally related genes together.

One of most fundamental and well-explored questions is to distinguish cancer driving mutations from passenger mutations (Fig. 7.1a). In Sect. 7.2, we discuss

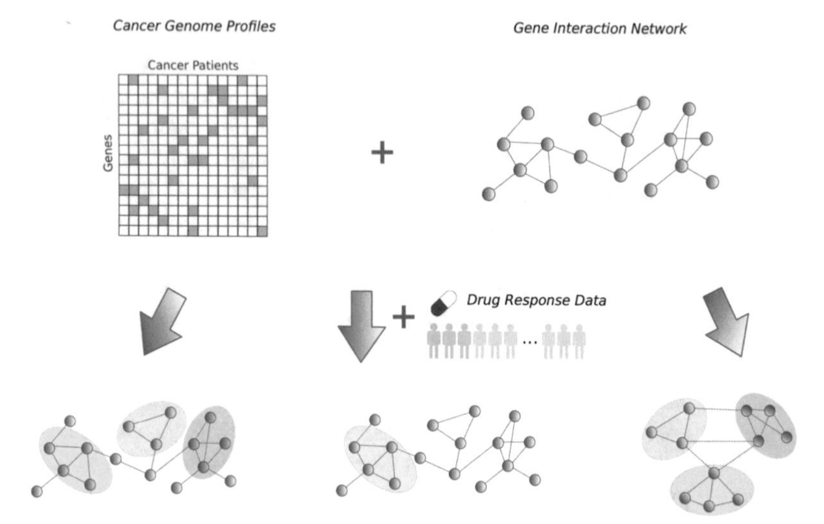

Fig. 7.1 Sophisticated computational tools need to be developed for large-scale cancer genome analyses. Several fundamental questions in cancer have been considered including (**a**) cancer driver module identification, (**b**) uncovering genetic causes of differences in drug response or other continuous cancer phenotypes, and (**c**) understanding complex relationships between genes and their alterations in module level

how methods based on graph theory and combinatorial optimization can be used to identify subnetworks containing cancer drivers. In Sect. 7.3, we consider additional data with continuous cancer phenotypes such as drug response and present algorithms to identify subnetworks associated specifically with the given phenotype (Fig. 7.1b). Another interesting problem is to identify cancer-related gene modules interacting with each other in pathway levels. In Sect. 7.4, we discuss algorithms to uncover various relationships among cancer gene modules based on functional interactions and mutation profiles (Fig. 7.1c).

7.2 Identifying Cancer Driver Genes and Mutations

Cancer genomes contain mutations accumulated over time, and while some are cancer driving mutations that confer cancer cell growth, most of them are benign passenger mutations. Distinguishing cancer driving mutations from passenger mutations is important but challenging due to the heterogeneity of cancer mutations. The set cover approach, a classical combinatorial optimization problem, is found to be useful to overcome the challenges arising from cancer heterogeneity [6, 19]. Integrated with network information, the approach has been further extended to identify cancer driving pathways [7, 17, 20, 21, 35]. By identifying a connected

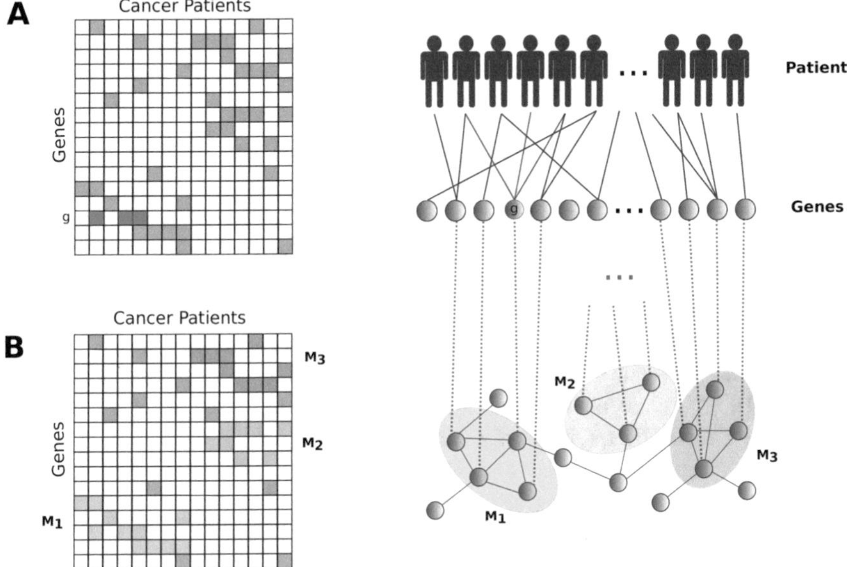

Fig. 7.2 Module Cover algorithm. Cancer driver identification via connected set cover algorithm. (**a**) Given a mutation table M, a set cover instance can be constructed by defining a gene g covering patients if they have mutations in g, and (**b**) adding interaction information among genes, *Module Cover* algorithm identifies a set of connected subnetworks covering patients

set of mutated genes that are functionally related, the methods can uncover cancer causing pathways in the presence of heterogeneous mutations. Alternatively, network propagation-based approaches have been also considered for cancer driver pathway discovery (for a review of propagation methods, see [9] and references therein).

Set Cover Approach A set cover is a classical problem in combinatorial optimization defined as follows: given a set P of elements and a collection of sets $S = \{S_1, S_2, ..\}$, where $S_i \subset P$, find $S' \subseteq S$ with minimum $|S'|$ to cover all elements in P (i.e., $\bigcup_{S_i \in S'} S_i = P$). For a given cancer mutation profile M, one can construct a set cover instance as follows (Fig. 7.2a): patients are considered as elements and a gene g is covering a patient p if and only if the gene g is mutated in the patient p (i.e., there is an edge between g and p in the bipartite graph shown in Fig. 7.2a). In other words, the set that gene g covering is defined as the set of patients having mutations in the gene g. The goal is to select the smallest set of genes covering (almost) all the patients. Note that by covering all patients, the mutations of selected gene set explain all cancers, while the solution allows to select different mutations for different patients, taking into account the heterogeneity of cancer mutations. Several variants of set cover problem can be considered depending on applications. For example, the weighted version of the problem allows each set to have its associated weights, and the objective is to minimize the total weight of the selected sets. In another application, in which the number of genes to be selected is given, the objective can be to cover as many elements as possible. Most of set cover variants are NP-hard, but greedy or integer linear program (ILP) approaches have been used to solve the problems [17, 20].

Connected Set Cover Set cover approaches can be extended by combining the mutation (or differential expression) status M of genes with the interaction information $G = (V, E)$, where V is a set of genes and E represents interactions (Fig. 7.2b). We assign weights w_e to the edges based on interaction confidence scores or mutation patterns (see more discussion below). *Module Cover* algorithm was developed with the goal of identifying connected subnetworks of genes whose mutations or differential expression collectively covers cancer genomes [20]. Formally, *Module Cover* is to select a set of gene modules, $GM = \{GM_1, GM_2, \ldots\}$, where $GM_i \subset V$ so as to cover all the patients and minimize the total cost $\sum_{GM_i \in GM} COST(GM_i)$, where $COST(GM_i)$ is defined as

$$|GM_i| - \sum_{g_1 \in GM_i} \sum_{g_2 \in GM_i, g_1 \neq g_2} w_{(g_1,g_2)}/(|GM_i| - 1), \tag{7.1}$$

where $w_{(g_1,g_2)}$ is the weight of edge (g_1, g_2). That is, the algorithm minimizes the number of selected genes minus the sum of the average edge weights of genes within the same modules.

Since the problem is NP-hard, *Module Cover* selects a gene greedily in each iteration by choosing the one that maximizes the ratio of benefit and cost. The algorithm also decides, based on the cost, whether to include the selected gene in one of existing modules or create a new module. Formally, the algorithm works as follows: at each l-th iteration, $BENEFIT_g(l)$ is computed as $S_g \setminus COV(l-1)$, where S_g is the set of patients whom g covers and $COV(l-1)$ is the set of all patients covered at $(l-1)$-th iteration. The cost of adding g is obtained by $COST_g(l) = min(1, min_{GM_i \in GM(l-1)}(COST(GM_i \bigcup \{g\}) - COST(GM_i)))$. $GM(l-1)$ is the set of modules selected by $(l-1)$-th iteration. At l-iteration, the gene with $BENEFIT_g(l)/COST_g(l)$ is selected. If the minimum cost is 1, then g creates a separate module. Otherwise, the gene is added to the module that incurs the minimum cost increase.

As shown in Fig. 7.2b, the algorithm finds a collection of subnetworks whose mutations cover or explain cancer patients but each patient may be covered by a different set of modules. Since it is known that cancer development requires multiple mutated genes (or pathways), the algorithm can also take another parameter k, which indicates by how many genes a patient should be covered.

The connected set cover problem was considered in several other applications and using different approaches [7, 17, 21, 35]. For example, Ulitsky et al. developed a computational method for identifying a connected set of genes with differential expression covering all patients and applied the algorithm to analyze seven human diseases, including Parkinson's disease [35]. *nCOP* aims to find a "small" connected subnetwork of genes to cover "most" patients with a trade-off parameter α to balance two competing objectives—the size of the network and the number of covered patients [17]. Two algorithms, one based on integer linear program (ILP) and another based on a fast greedy heuristic were considered for cancer driver identification with pan-cancer data across 24 different cancer types. The constraints using a commodity flow technique were added to ILP to ensure connectivity of genes. Note that although obtaining the optimal solution of an integer linear program instance is in general NP-hard, the solutions in many applications can be obtained in a reasonable amount time using an optimization package such as CPLEX [10].

Augmenting Edge Weights with Mutual Exclusivity Different types of interaction networks and the weights of nodes and edges can be used to improve the results. In *MEMCover*, an extension of *Module Cover* algorithm, mutual exclusivity scores were incorporated into edge weights. As discussed in the introduction, mutual exclusivity can arise due to different reasons. Several approaches have been considered to measure the significance of mutual exclusivity or classify the types of mutual exclusivity. Permutation tests were commonly used to compute p-values, in which a given mutation matrix is permuted while preserving the mutation frequencies of genes and patients.

For pan-cancer analysis where thousands of cancer genomes from multiple tissue types were analyzed together, *MEMCover* used a *type restricted* permutation test to measure the significance of mutual exclusivity of cancer mutations. In a type restricted permutation, permutations were performed in each tissue type separately and the overall mutual exclusivity scores are computed by merging the scores across tissue types. By doing so, the gene pairs that are mutually exclusive within each tissue type for multiple tissue types commonly will be captured. We hypothesize that those genes are more likely to be common cancer driver genes across multiple tissues. The mutual exclusivity scores were subsequently combined with functional interaction scores to compute edge weights. It is shown that using mutual exclusivity scores improves cancer driver identification. Applied to analyze TCGA Pan-Cancer dataset, *MEMCover* identified a large number of subnetworks enriched with genes in cancer-related pathways as well as novel subnetworks whose role across cancer types has been less appreciated.

7.3 Phenotype-Associated Mutations

Cancer heterogeneity extends to other phenotypes such as drug response, survival time, or mutational signatures. While most of the module identification methods in cancer have been focused on finding cancer drivers, one important problem in this area is to understand genetic causes of differences in continuous cancer phenotypes. For example, cancer patients even in the same cancer subtypes respond to a drug differently, and it is critical to predict drug responses on a personalized level for cancer treatment. Phenotypic heterogeneity in cancer is often caused by different patterns of genetic alterations. Understanding such phenotype–genotype relationships is fundamental for the advance of personalized medicine. Genome-wide association approaches may be used to partly answer the question. However, such approaches do not fully utilize the complex proprieties of cancer mutational landscape and interactions such as mutual exclusivity. Recognizing the need to consider cancer heterogeneity and module level association of mutations with a phenotype, algorithms have been developed to find mutated gene modules associated with a phenotype [6, 18, 22, 24, 26]. Below we start with a set cover-based approach that accommodates associations with a continuous phenotype. We then discuss how the idea can be further extended to identify subnetworks associated with a phenotype. The problem is formulated as an integer linear program and solved optimally to obtain a set of genes satisfying the constraints.

Set Cover Approach *UNCOVER* is developed to find a set of genes whose mutations are associated with a continuous cancer phenotype. In addition to a binary mutation matrix M that provides the alteration status for genes, the input of *UNCOVER* includes a continuous phenotype vector $W = \{w_1, w_2, \ldots, w_N\}$ for the same set of N cancer patients. The objective is to select a set of k genes that maximizes the total weight of covered patients by the selected genes. In addition, to preferentially select mutually exclusive genes, *UNCOVER* adds a penalty for overlapping mutations.

Let a collection of set $S = \{S_1, S_2, \ldots\}$ defined based on alteration table M as described in Sect. 7.2. S_i is the set of patients covered by (or having mutations in) gene g_i. Given a selected collection of set $S' = \{S_{i_1}, S_{i_2}, \ldots S_{i_k}\}$ of size K, the objective of *UNCOVER* is to maximize

$$\sum_{j \in C} w_j - \sum_{j \in C} (c_j - 1) p_j, \tag{7.2}$$

where C is the set of patients who belong to any set in S' (i.e., $C = \bigcup_{S_i \in S'} S_i$) and c_j is the number of times a patient j is covered by S' (i.e., $|\{S_i \in S' | j \in S_i\}|$) and p_j is the penalty for overlapping mutations (when $c_j > 1$). The objective of *UNCOVER* is to select K genes that minimizes the cost given in Eq. (7.2).

Two algorithms were proposed to find solutions; one based on integer linear program to find the optimal solution and a greedy algorithm for fast computation. The greedy algorithm picks the gene that produces the maximum increase of the objective value in each iteration. The significance of identified modules was measured with a permutation test by permuting the phenotype vector W.

Connected Set Cover The ideas can be extended to find functional modules associated with a phenotype by considering gene interaction information together. *NetPhix* (NETwork-to-PHenotpe association with eXclusivity) aims to identify mutated subnetworks that are associated with continuous cancer phenotypes (Fig. 7.3). Similar to the methods for identifying driver pathways, we utilize a connected set cover approach to identify subnetworks of genes that have heterogeneous mutation patterns but are functionally related. However, different from finding cancer driver

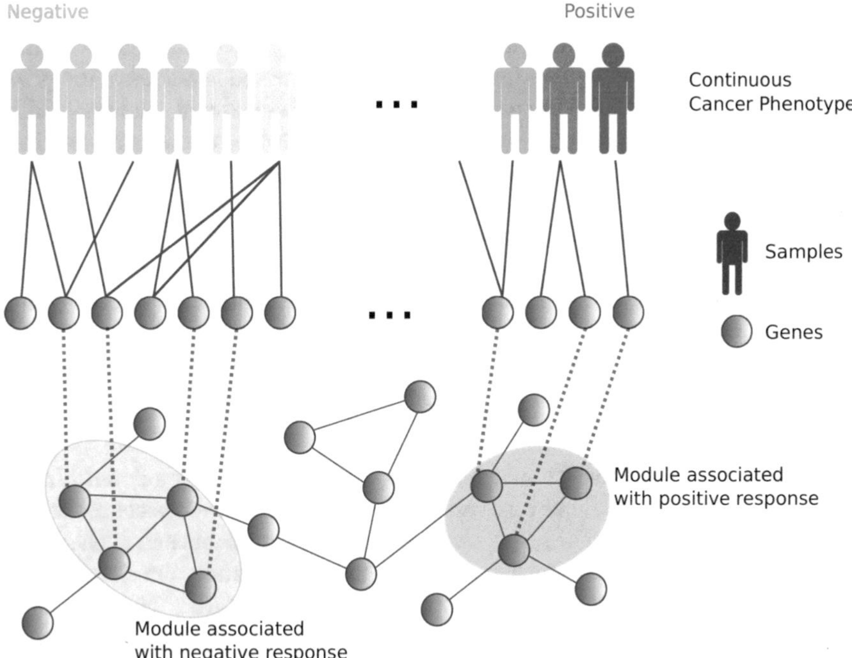

Fig. 7.3 Connected set cover for phenotype-associated networks. To identify phenotype-associated networks, *NetPhix* utilizes phenotype data for patients such as drug response information in addition to mutation matrix *M*

genes based on gene alterations, the goal here is to find a connected set of genes whose alterations are *associated with phenotyphic differences*. Utilizing network information, *NetPhix* can identify functionally coherent subnetworks associated with continuous cancer phenotypes.

While *Module Cover* takes a greedy strategy to find a solution to the connected set cover problem, *NetPhix* formulates the problem as an integer linear program (ILP) to maximize the objective function given in Eq. (7.2), which can be solved optimally using optimization packages such as CPLEX. There are a couple of ways to enforce connectivity of the identified networks in ILP and *NetPhix* uses the density constraints to ensure that each gene is connected to a certain fraction δ of genes in the module and by setting $\delta \geq 0.5$, the identified subnetwork is guaranteed to be connected. Other connectivity constraint approaches include using a branch-and-cut algorithm [14] or commodity flow technique [17].

Uncovering Mutated Networks Associated with Drug Responses One application of *NetPhix* is to identify a subnetwork of genes whose mutations are associated with drug response. It is a critical problem in cancer treatment to effectively predict drug responses and understand the genetic causes of the differences in responses. Due to the heterogeneity of cancer, the response of a drug for patients can be varied even in the same cancer subtypes. The different responses may be related to different mutation status of genes for the patients. Large-scale drug screening datasets such as GDSC (Genomics of Drug Sensitivity in Cancer) [18] tested hundreds of drugs for thousands of cell lines, and the cell survival rates were measured. The availability of such drug screening results together with genomic profiles of the same cell lines made it possible to systematically analyze drug responses in a personalized level.

NetPhix can be applied for the analysis of such datasets to better predict the responses and identify the genomic causes for differences in the response. Importantly, the drug responses are given as continuous values in the datasets such as IC_{50} or AUC values. *NetPhix* can fully utilize the continuous values without binarization for better prediction. In addition, using network information helps to find a functionally coherent set of genes whose mutations are associated with drug response. Applied for the analysis of GDSC dataset, *NetPhix* identified gene modules significantly associated with many drugs.

Understanding the Etiology of Mutational Signatures Utility of *NetPhix* can be extended to finding subnetworks associated with any continuous cancer phenotypes. In another application, *NetPhix* was used to uncover mutated subnetworks associated with mutational signatures [25]. Mutational signatures are characteristic mutational patterns that were produced by different mutagenic processes such as carcinogenic exposures or cancer-related aberrations of DNA maintenance machinery.

Understanding the etiology of mutational signatures affecting a cancer genome can be instrumental for understanding cancer progression and designing personalized cancer therapy. One interesting question is if there is a pathway whose alterations might have caused specific mutational signatures. If the mutational signature was caused by an endogenous source such as a malfunction of cancer-related pathways and is associated with mutations in the pathway, *NetPhix* can be applied to uncover the subnetwork underlying the specific mutational signature.

Computational methods to decompose mutations in cancer genomes can output the individual strength of each mutational signature for each patient [32, 39]. For each signature j, we are given a vector $ms(j) = \{ms_1^j, ms_2^j, \ldots, ms_N^j\}$, where ms_i^j indicates the strength of mutational signature j for patient i. To obtain (if any) a subnetwork whose mutations might be associated with the mutational signature j, we can run *NetPhix* with mutation matrix M and signature strength vector $ms(j)$ for signature j as inputs. Our analysis with the mutational signatures produced by *SigMa* [39] provides novel insights into mutagenic processes in breast cancer. In particular, the results demonstrate important differences in the etiology of APOBEC-related signatures (COSMIC Signature 2 and 13 [13]) and identify different subnetworks for clustered and dispersed mutations. These outcomes are important for understanding mutagenic processes in cancer and for developing personalized drug therapies.

7.4 Interactions Between Modules

As discussed in the introduction, mutations in two genes may exhibit various patterns such as mutual exclusivity and co-occurrences. Such relationships between genes may help identify important properties in cancer. As dysregulated pathways can have different causal effects in cancer, one can also think of such relationships existing between modules. Given a variety of reasons that mutational patterns arise, analyzing those relationships in genes- and modules-level simultaneously can help understand the disease. Such algorithms seek multiple modules that have properties both within a module and between modules. Identifying interactions between modules is more challenging due to the exponential increase of possible combinations.

Modules interacting with each other have been considered in the context of differential expression patterns (correlated genes within a module and anti-correlation between modules), genetic interactions (alleviating or aggravating), or different cancer subtypes [1, 2, 12, 16]. It is also closely related to correlation clustering, a well-studied combinatorial optimization problem in various clustering applications [5].

BeWith Method Methods for joint analysis of different relationships can lead to a better understanding of the causes and impacts of cancer mutations. *BeWith* (BEtween-WITHin method) is a method developed to identify groups of genes (or gene modules) with coherent patterns within modules but distinct properties between genes in different modules.

The input of *BeWith* algorithm [12] is a weighted graph $G = (V, E)$ with a set of genes V, and edges E (Fig. 7.4a). Two types of edge weights w_{be} and w_{with} are given. The between edge weights w_{be} represent the relationships between modules (when two genes belong to two different modules), and the within edge weights w_{with} are for gene pairs in the same module (when the two genes are in the same module). Depending on applications, both weights can be defined based on, for example, mutual exclusivity, co-occurrences, and functional interactions. Mutual exclusivity and/or co-occurrences of mutations can be computed from mutation profile M, and there are several methods to compute the scores efficiently [23, 31].

The objective of *BeWith* is to extract a set of K modules that maximizes the sum of between weights and within weights, given parameter K for the number of modules and m for the maximum number of genes in each module (Fig. 7.4b). Formally, *BeWith* selects a set of gene modules $GM = \{GM_1, GM_2, \dots GM_K\}$, where $GM_i \subset V$ so as to maximize

$$\sum_{u \in GM_i, v \in GM_j, i \neq j} w_{be}(u, v) + \sum_{i=1}^{K} \sum_{u,v \in GM_i} w_{with}(u, v). \qquad (7.3)$$

The problem is formulated as an integer linear program and solved optimally. For connectivity, we make sure that the genes inside each module is connected by

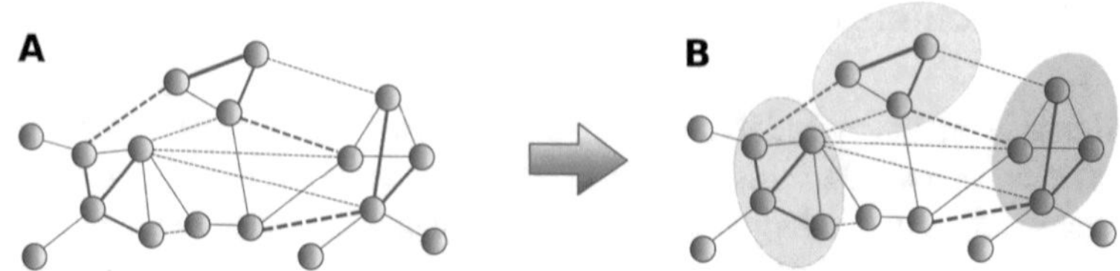

Fig. 7.4 BeWith algorithm. BeWith identifies a set of gene modules, which are coherent inside a module and distinct from different modules. (**a**) Two types of edge weights—weights between modules w_{be} (red dotted lines) and within modules (black solid lines) w_{with} are given. (**b**) BeWith finds three modules to maximize the sum of total weight between modules (w_{be}) and total weight within modules (w_{with})

adding the ILP constraints that each gene is connected to a certain fraction δ of genes in the same module ($\delta \geq 0.5$).

BeWith is a general framework, and by defining edge weights in different ways, one can investigate various aspects of cancer mutational patterns, leading to uncover relationships between mutated gene modules, cancer subtypes, and mutational signatures. Dao et al. [12] considered three different settings of BeWith for the analysis of cancer mutations: (1) *BeME-WithFun*, in which the relations between modules are enriched with mutual exclusivity, while genes within each module are functionally related; (2) *BeME-WithCo*, which combines mutual exclusivity between modules with co-occurrence within modules; and (3) *BeCo-WithMEFun*, which ensures co-occurrence between modules, while the within module relations combine mutual exclusivity and functional interactions. Below, we discuss the first setting (*BeME-WithFun*) in more detail as an example.

Identifying Functional Modules for Different Cancer Subtypes BeWith can be utilized to find a set of functional driver modules that belong to different cancer subtypes. That is, the goal is to find multiple modules such that genes in each module are functionally coherent (i.e., in the same pathway), but mutations in different modules are associated with different set of cancer patients (i.e., belong to different cancer subtypes). For this purpose, we use the setting named *BeMEWithFun* by defining the between edge weights based on mutual exclusivity but without functional interaction (*BeME*) and within edge weights based on their functional interaction scores (*WithFun*). It is shown that the mutual exclusivity "without" functional interactions is more likely to indicate that the mutated genes are causal to different cancer subtypes [21], while the mutual exclusivity between functionally related genes may indicate the genes are driver genes in the same pathways. By applying the setting to TCGA breast cancer (BRCA) and endometrial cancer (UCEC) datasets [38], we uncovered different modules significantly enriched in different cancer subtypes [12]. The general framework of BeWith makes it possible to apply the algorithm in a broad range of applications in which between and within relationships are given.

7.5 Conclusion

Cancer is a disease caused by dyregulated pathways rather than a single gene mutation. Cancer manifests heterogeneity in multiple aspects, and network-based approaches have been shown to provide effective solutions for cancer analysis by considering mutated genes in functional context. In this chapter, we considered three questions critical to advance our understanding of cancer mutations: (1) identifying cancer driving pathways, (2) uncovering subnetworks associated with continuous cancer phenotypes such as drug response, and (3) extracting gene modules that are interacting with each other. In all problems, combining network information helps identify modules that are more functionally coherent and therefore easier to interpret.

Acknowledgment This work was supported by the Intramural Research Program of the National Institutes of Health, National Library of Medicine.

References

1. Amar, D., Shamir, R.: Constructing module maps for integrated analysis of heterogeneous biological networks. Nucleic Acids Res. **42**(7), 4208–4219 (2014)
2. Amar, D., Safer, H., Shamir, R.: Dissection of regulatory networks that are altered in disease via differential co-expression. PLoS Comput. Biol. **9**(3), e1002955 (2013)
3. J. Armenia, S.A.M. Wankowicz, D. Liu, J. Gao, R. Kundra, E. Reznik, W.K. Chatila, D. Chakravarty, G.C. Han, I. Coleman, B. Montgomery, C. Pritchard, C. Morrissey, C.E. Barbieri, H. Beltran, A. Sboner, Z. Zafeiriou, S. Miranda, C.M. Bielski, A.V. Penson, C. Tolonen, F.W. Huang, D. Robinson, Y.M. Wu, R. Lonigro, L.A. Garraway, F. Demichelis, P.W. Kantoff, M.E. Taplin, W. Abida, B.S. Taylor, H.I. Scher, P.S. Nelson, J.S. de Bono, M.A. Rubin, C.L. Sawyers, A.M. Chinnaiyan, N. Schultz, E.M. Van Allen, The long tail of oncogenic drivers in prostate cancer. Nat. Genet. **50**(5), 645–651 (2018)
4. Ö. Babur, M. Gönen, B.A. Aksoy, N. Schultz, G. Ciriello, C. Sander, E. Demir, Systematic identification of cancer driving signaling pathways based on mutual exclusivity of genomic alterations. Genome Biol. **16**(1), 45 (2015)
5. N. Bansal, A. Blum, S. Chawla, Correlation clustering. Mach. Learn. **56**(1–3), 89–113 (2004)
6. R.S. Basso, D.S. Hochbaum, F. Vandin, Efficient algorithms to discover alterations with complementary functional association in cancer. In: Research in Computational Molecular Biology (RECOMB) 2018, pp. 278–279 (2018)
7. S.A. Chowdhury, M. Koyuturk, Identification of coordinately dysregulated subnetworks in complex phenotypes. Pac. Symp. Biocomput. **15**, 133–144 (2010)
8. Ciriello, G., Cerami, E., Aksoy, B.A., Sander, C., Schultz, N.: Using MEMo to discover mutual exclusivity modules in cancer. Curr. Protoc. Bioinform. (2013); Chapter 8: Unit 8.17. https://doi.org/10.1002/0471250953.bi0817s41
9. Cowen, L., Ideker, T., Raphael, B.J., Sharan, R.: Network propagation: a universal amplifier of genetic associations. Nat. Rev. Genet. **18**(9), 551–562 (2017)
10. CPLEX. https://www.ibm.com/analytics/cplex-optimizer
11. Cristea, S., Kuipers, J., Beerenwinkel, N.: pathTiMEx: joint inference of mutually exclusive cancer pathways and their progression dynamics. J. Comput. Biol. **24**(6), 603–615 (2017)

12. Dao, P., Kim, Y.A., Wojtowicz, D., Madan, S., Sharan, R., Przytycka, T.M.: BeWith: a Between-Within method to discover relationships between cancer modules via integrated analysis of mutual exclusivity, co-occurrence and functional interactions. PLoS Comput. Biol. **13**(10), e1005695 (2017)
13. Forbes, S.A., Beare, D., Boutselakis, H., Bamford, S., Bindal, N., et al.: Cosmic: somatic cancer genetics at high-resolution. Nucl. Acids Res. **45**(D1), D777–D783 (2017)
14. Frith, M.C., Pedersen, C.N.S. (eds.) Algorithms in Bioinformatics - 16th International Workshop, WABI 2016, Aarhus, August 22–24, 2016. Proceedings. Lecture Notes in Computer Science, vol. 9838. Springer, New York (2016)
15. Hofree, M., Shen, J.P., Carter, H., Gross, A., Ideker, T.: Network-based stratification of tumor mutations. Nat. Methods **10**(11), 1108–1115 (2013)
16. Hou, J.P., Emad, A., Puleo, G.J., Ma, J., Milenkovic, O.: A new correlation clustering method for cancer mutation analysis. Bioinformatics **32**(24), 3717–3728 (2016)
17. Hristov, B.H., Singh, M.: Network-based coverage of mutational profiles reveals cancer genes. Cell Syst. **5**(3), 221–229 (2017)
18. Iorio, F., Knijnenburg, T.A., Vis, D.J., Bignell, G.R., Menden, M.P., Schubert, M., Aben, N., Goncalves, E., Barthorpe, S., Lightfoot, H., Cokelaer, T., Greninger, P., van Dyk, E., Chang, H., de Silva, H., Heyn, H., Deng, X., Egan, R.K., Liu, Q., Mironenko, T., Mitropoulos, X., Richardson, L., Wang, J., Zhang, T., Moran, S., Sayols, S., Soleimani, M., Tamborero, D., Lopez-Bigas, N., Ross-Macdonald, P., Esteller, M., Gray, N.S., Haber, D.A., Stratton, M.R., Benes, C.H., Wessels, L.F.A., Saez-Rodriguez, J., McDermott, U., Garnett, M.J.: A landscape of pharmacogenomic interactions in cancer. Cell **166**(3), 740–754 (2016)
19. Kim, Y.A., Wuchty, S., Przytycka, T.M.: Identifying causal genes and dysregulated pathways in complex diseases. PLoS Comput. Biol. **7**(3), e1001095 (2011)
20. Kim, Y.A., Salari, R., Wuchty, S., Przytycka, T.M.: Module cover - a new approach to genotype-phenotype studies. Pac. Symp. Biocomput. **15**, 135–146 (2013)
21. Kim, Y.A., Cho, D.Y., Dao, P., Przytycka, T.M.: MEMCover: integrated analysis of mutual exclusivity and functional network reveals dysregulated pathways across multiple cancer types. Bioinformatics **31**(12), i284–292 (2015)
22. Kim, J.W., Botvinnik, O.B., Abudayyeh, O., Birger, C., Rosenbluh, J., Shrestha, Y., Abazeed, M.E., Hammerman, P.S., DiCara, D., Konieczkowski, D.J., Johannessen, C.M., Liberzon, A., Alizad-Rahvar, A.R., Alexe, G., Aguirre, A., Ghandi, M., Greulich, H., Vazquez, F., Weir, B.A., Van Allen, E.M., Tsherniak, A., Shao, D.D., Zack, T.I., Noble, M., Getz, G., Beroukhim, R., Garraway, L.A., Ardakani, M., Romualdi, C., Sales, G., Barbie, D.A., Boehm, J.S., Hahn, W.C., Mesirov, J.P., Tamayo, P.: Characterizing genomic alterations in cancer by complementary functional associations. Nat. Biotechnol. **34**(5), 539–546 (2016)
23. Kim, Y.-A., Madan, S., Przytycka, T.M.: WeSME: uncovering mutual exclusivity of cancer drivers and beyond. Bioinformatics **33**(6), 814–821 (2017). https://doi.org/10.1093/bioinformatics/btw242
24. Kim, Y.A., Sarto Basso, R., Wojtowicz, D., Liu, A.S., Hochbaum, D.S., Vandin, F., Przytycka, T.M.: Identifying drug sensitivity subnetworks with NETPHIX. iScience, **23**(10), 101619 (2020)
25. Kim, Y.A., Wojtowicz, D., Sarto Basso, R., Sason, I., Robinson, W., Hochbaum, D.S., Leiserson, M.D.M., Sharan, R., Vadin, F., Przytycka, T.M.: Network-based approaches elucidate differences within APOBEC and clock-like signatures in breast cancer. Genome Med. **12**(1), 52 (2020)

    Kim, Y.-A., Wojtowicz, D., Basso, R.S., Sason, I., Robinson, W., Hochbaum, D.S., Leiserson, M.D.M., Sharan, R., Vandin, F., Przytycka, T.M.: Network-based approaches elucidate differences within APOBEC and clock-like signatures in breast cancer. Genome Med. 12, 1:52 (2020)
26. Knijnenburg, T.A., Klau, G.W., Iorio, F., Garnett, M.J., McDermott, U., Shmulevich, I., Wessels, L.F.: Logic models to predict continuous outputs based on binary inputs with an application to personalized cancer therapy. Sci. Rep. **6**, 36812 (2016)

27. Lawrence, M.S., Stojanov, P., Mermel, C.H., Robinson, J.T., Garraway, L.A., Golub, T.R., Meyerson, M., Gabriel, S.B., Lander, E. S, Getz, G.: Discovery and saturation analysis of cancer genes across 21 tumour types. Nature **505**(7484), 495–501 (2014)
28. Leiserson, M.D.M., Blokh, D., Sharan, R., Raphael, B.J.: Simultaneous identification of multiple driver pathways in cancer. PLoS Comput. Biol. **9**(5), e1003054 (2013)
29. Leiserson, M.D.M., Vandin, F., Wu, H.-T., Dobson, J.R., Eldridge, J.V., Thomas, J.L., Papoutsaki, A., Kim, Y., Niu, B., McLellan, M., Lawrence, M.S., Gonzalez-Perez, A., Tamborero, D., Cheng, Y., Ryslik, G.A., Lopez-Bigas, N., Getz, G., Ding, L., Raphael, B.J.: Pan-cancer network analysis identifies combinations of rare somatic mutations across pathways and protein complexes. Nat. Genet. **47**(2), 106–114 (2015)
30. Leiserson, M.D.M., Wu, H.-T., Vandin, F., Raphael, B.J.: Comet: a statistical approach to identify combinations of mutually exclusive alterations in cancer. Genome Biol. **16**(1), 160 (2015)
31. Leiserson, M.D.M., Reyna, M.A., Raphael, B.J.: A weighted exact test for mutually exclusive mutations in cancer. Bioinformatics **32**(17), i736–i745 (2016)
32. Nik-Zainal, S., Davies, H., Staaf, J., Ramakrishna, M., Glodzik, D., et al.: Landscape of somatic mutations in 560 breast cancer whole-genome sequences. Nature **534**(7605), 47–54 (2016)
33. Silverbush, D., Cristea, S., Yanovich-Arad, G., Geiger, T., Beerenwinkel, N., Sharan, R.: Simultaneous integration of multi-omics data improves the identification of cancer driver modules. Cell Syst. **8**(5), 456–466 (2019)
34. Szczurek, E., Beerenwinkel, N.: Modeling mutual exclusivity of cancer mutations. PLoS Comput. Biol. **10**(3), e1003503 (2014)
35. Ulitsky, I., Krishnamurthy, A., Karp, R.M., Shamir, R.: DEGAS: de novo discovery of dysregulated pathways in human diseases. PLoS One **5**(10), e13367 (2010)
36. Vandin, F., Upfal, E., Raphael, B.J.: De novo discovery of mutated driver pathways in cancer. Genome Res. **22**(2), 375–385 (2012)
37. Vandin, F., Upfal, E., Raphael, B.J.: De novo discovery of mutated driver pathways in cancer. Genome Res. **22**(2), 375–385 (2012)
38. Weinstein, J.N., Collisson, E.A., Mills, G.B., Shaw, K.R., Ozenberger, B.A., Ellrott, K., Shmulevich, I., Sander, C., Stuart, J.M., Chang, K., Creighton, C.J., Davis, C., Donehower, L.L., Drummond, J., Wheeler, D., Ally, A., Balasundaram, M., Birol, I., Butterfield, S.N., Chu, A., Chuah, E., Chun, H.J., Dhalla, N., Guin, R., Hirst, M., Hirst, C., Holt, R.A., Jones, S.J., Lee, D., Li, H.I., Marra, M.A., Mayo, M., Moore, R.A., Mungall, A.J., Robertson, A.G., Schein, J.E., Sipahimalani, P., Tam, A., Thiessen, N., Varhol, R.J., Beroukhim, R., Bhatt, A.S., Brooks, A.N., Cherniack, A.D., Freeman, S.S., Gabriel, S.B., Helman, E., Jung, J., Meyerson, M., Ojesina, A.I., Pedamallu, C.S., Saksena, G., Schumacher, S.E., Tabak, B., Zack, T., Lander, E.S., Bristow, C.A., Hadjipanayis, A., Haseley, P., Kucherlapati, R., Lee, S., Lee, E., Luquette, L.J., Mahadeshwar, H.S., Pantazi, A., Parfenov, M., Park, P.J., Protopopov, A., Ren, X., Santoso, N., Seidman, J., Seth, S., Song, X., Tang, J., Xi, R., Xu, A.W., Yang, L., Zeng, D., Auman, J.T., Balu, S., Buda, E., Fan, C., Hoadley, K.A., Jones, C.D., Meng, S., Mieczkowski, P.A., Parker, J.S., Perou, C.M., Roach, J., Shi, Y., Silva, G.O., Tan, D., Veluvolu, U., Waring, S., Wilkerson, M.D., Wu, J., Zhao, W., Bodenheimer, T., Hayes, D.N., Hoyle, A.P., Jeffreys, S.R., Mose, L.E., Simons, J.V., Soloway, M.G., Baylin, S.B., Berman, B.P., Bootwalla, M.S., Danilova, L., Herman, J.G., Hinoue, T., Laird, P.W., Rhie, S.K., Shen, H., Triche, T., Weisenberger, D.J., Carter, S.L., Cibulskis, K., Chin, L., Zhang, J., Getz, G., Sougnez, C., Wang, M., Saksena, G., Carter, S.L., Cibulskis, K., Chin, L., Zhang, J., Getz, G., Dinh, H., Doddapaneni, H.V., Gibbs, R., Gunaratne, P., Han, Y., Kalra, D., Kovar, C., Lewis, L., Morgan, M., Morton, D., Muzny, D., Reid, J., Xi, L., Cho, J., DiCara, D., Frazer, S., Gehlenborg, N., Heiman, D.I., Kim, J., Lawrence, M.S., Lin, P., Liu, Y., Noble, M.S., Stojanov, P., Voet, D., Zhang, H., Zou, L., Stewart, C., Bernard, B., Bressler, R., Eakin, A., Iype, L., Knijnenburg, T., Kramer, R., Kreisberg, R., Leinonen, K., Lin, J., Liu, Y., Miller, M., Reynolds, S.M., Rovira, H., Shmulevich, I., Thorsson, V., Yang, D., Zhang, W., Amin, S., Wu, C.J., Wu, C.C., Akbani, R., Aldape, K., Baggerly, K.A., Broom, B., Casasent, T.D., Cleland, J., Creighton, C., Dodda, D., Edgerton, M., Han, L., Herbrich, S.M., Ju, Z., Kim, H., Lerner, S., Li, J., Liang, H., Liu,

W., Lorenzi, P.L., Lu, Y., Melott, J., Mills, G.B., Nguyen, L., Su, X., Verhaak, R., Wang, W., Weinstein, J.N., Wong, A., Yang, Y., Yao, J., Yao, R., Yoshihara, K., Yuan, Y., Yung, A.K., Zhang, N., Zheng, S., Ryan, M., Kane, D.W., Aksoy, B.A., Ciriello, G., Dresdner, G., Gao, J., Gross, B., Jacobsen, A., Kahles, A., Ladanyi, M., Lee, W., Lehmann, K.V., Miller, M.L., Ramirez, R., Ratsch, G., Reva, B., Sander, C., Schultz, N., Senbabaoglu, Y., Shen, R., Sinha, R., Sumer, S.O., Sun, Y., Taylor, B.S., Weinhold, N., Fei, S., Spellman, P., Benz, C., Carlin, D., Cline, M., Craft, B., Ellrott, K., Goldman, M., Haussler, D., Ma, S., Ng, S., Paull, E., Radenbaugh, A., Salama, S., Sokolov, A., Stuart, J.M., Swatloski, T., Uzunangelov, V., Waltman, P., Yau, C., Zhu, J., Hamilton, S.R., Getz, G., Sougnez, C., Abbott, S., Abbott, R., Dees, N.D., Delehaunty, K., Ding, L., Dooling, D.J., Eldred, J.M., Fronick, C.C., Fulton, R., Fulton, L.L., Kalicki-Veizer, J., Kanchi, K.L., Kandoth, C., Koboldt, D.C., Larson, D.E., Ley, T.J., Lin, L., Lu, C., Magrini, V.J., Mardis, E.R., McLellan, M.D., McMichael, J.F., Miller, C.A., O'Laughlin, M., Pohl, C., Schmidt, H., Smith, S.M., Walker, J., Wallis, J.W., Wendl, M.C., Wilson, R.K., Wylie, T., Zhang, Q., Burton, R., Jensen, M.A., Kahn, A., Pihl, T., Pot, D., Wan, Y., Levine, D.A., Black, A.D., Bowen, J., Frick, J., Gastier-Foster, J.M., Harper, H.A., Helsel, C., Leraas, K.M., Lichtenberg, T.M., McAllister, C., Ramirez, N.C., Sharpe, S., Wise, L., Zmuda, E., Chanock, S.J., Davidsen, T., Demchok, J.A., Eley, G., Felau, I., Ozenberger, B.A., Sheth, M., Sofia, H., Staudt, L., Tarnuzzer, R., Wang, Z., Yang, L., Zhang, J., Omberg, L., Margolin, A., Raphael, B.J., Vandin, F.F., Wu, H.T., Leiserson, M.D., Benz, S.C., Vaske, C.J., Noushmehr, H., Knijnenburg, T., Wolf, D., Van 't Veer, L., Collisson, E.A., Anastassiou, D., Ou Yang, T.H., Lopez-Bigas, N., Gonzalez-Perez, A., Tamborero, D., Xia, Z., Li, W., Cho, D.Y., Przytycka, T., Hamilton, M., McGuire, S., Nelander, S., Johansson, P., Jornsten, R., Kling, T., Sanchez, J.: The Cancer Genome Atlas Pan-Cancer analysis project. Nat. Genet. **45**(10), 1113–1120 (2013)

39. Wojtowicz, D., Sason, I., Huang, X., Kim, Y.A., Leiserson, M.D.M., Przytycka, T.M., Sharan, R.: Hidden Markov models lead to higher resolution maps of mutation signature activity in cancer. Genome Med **11**(1), 49 (2019)

Part III
Network-Based Omics Data Analysis

Chapter 8
Integrated Network-Based Computational Analysis for Drug Development

Mijin Kwon, Soorin Yim, Gwangmin Kim, and Doheon Lee

Abstract Network-based in silico approaches have played important roles in drug development. Network analysis is advantageous to identifying both direct (close) and indirect (distant) relationships among biological entities comprehensively. There have been many efforts to construct biological networks that reflect real biological phenomena as the first step of network analysis. A recently introduced CODA network consists of hundreds of cell-specific networks connected to each other through intercellular relationships. Each cell-specific network is structured into multiple layers including molecule level, cellular function level, and disease level, all of which are interconnected to each other. CODA network allows detailed interpretations or simulations of biological phenomena ranging from the molecule level to the disease level. This chapter describes the CODA-based computational models and discusses their benefits and/or limitations of each approach to achieve their aims.

Keywords Biological networks · Network medicine · Network theory · Drug development · Drug repositioning

8.1 Introduction

The rapid development of high-throughput techniques helps us to determine various molecular interactions, and molecular networks can be constructed by integrating them. Biological networks vary by the types of interactions (edges) between

M. Kwon · S. Yim · G. Kim
Department of Bio and Brain Engineering, KAIST, Daejeon, Republic of Korea
e-mail: mijinkwon@kaist.ac.kr; lsl4497@kaist.ac.kr; gwang5386@kaist.ac.kr

D. Lee (✉)
Department of Bio and Brain Engineering, KAIST, Daejeon, Republic of Korea

Bio-Synergy Research Center, Daejeon, Republic of Korea
e-mail: dhlee@kaist.ac.kr

© Springer Nature Switzerland AG 2021

B.-J. Yoon, X. Qian (eds.), *Recent Advances in Biological Network Analysis*,
https://doi.org/10.1007/978-3-030-57173-3_8

bio-molecules (nodes); protein–protein interaction, metabolic, signaling or transcriptional regulatory networks. Biological networks include not only molecular networks but also diverse heterogeneous networks such as drug-target networks and disease similarity networks. Biological networks are beneficial to comprehensively identifying complex events or associations between biological entities. Therefore, biological networks play increasingly more important roles in biological studies.

In [1], the authors constructed a large-scale network called CODA. A disease is caused due to a malfunction of some bio-molecules which affects normal cellular functions. To interpret these complex biological activities, CODA is structured into multiple layers, including a molecule level, a cellular function level, and a disease level layer. The molecule level network consists of intracellular intercellular interactions. These three layers are connected to each other based on the relationships between bio-molecules (e.g. genes, proteins, metabolites), cellular functions, or diseases. Heterogeneous information from various resources was integrated using a standard language format, CODA-ML [2] to integrate detailed biological interaction information within or between layers.

In Sects. 8.2, 8.3, 8.4, 8.5, 8.6, and 8.7, we introduce six studies that used CODA networks to identify new therapeutic genes for a disease, to predict therapeutic efficacies of individual/combined drugs or natural herbs, or to analyze the interactions between drugs and/or hormones. For each study, we mainly emphasized on describing the methodologies related to network analysis and discussing the benefits and shortcomings of each approach. Finally, in Sect. 8.8, we draw general conclusions and propose topics for future research.

8.2 Prediction of Therapeutic Targets

Target identification is the very early stage of drug development which aims at finding a new target that will be targeted by a drug. Among various approaches to target identification, biological network analysis is advantageous for several reasons. Firstly, unlike experimental methods, network analysis is time- and cost-efficient, which allows network analysis to be performed at a genome-scale. Secondly, network analysis allows us to simulate the effects of perturbing a specific gene, thereby predicting not only the therapeutic effect but also unintended side effects altogether.

For example, in [3], authors compared the phosphorylation status of proteins from normal (reference) state, atrophic state, and a specific gene-inhibited state to test whether the gene can be a therapeutic target for muscle atrophy (Fig. 8.1). To predict therapeutic targets, authors hypothesized that genes whose inhibition can reverse the phosphorylation status of proteins from atrophic state to normal state can be the therapeutic targets for muscle atrophy.

To test their hypothesis, authors firstly constructed myocyte-specific phosphorylation network. The authors constructed a backbone network based on phosphorylation and dephosphorylation reactions collected from three databases [4–6]. Then, the

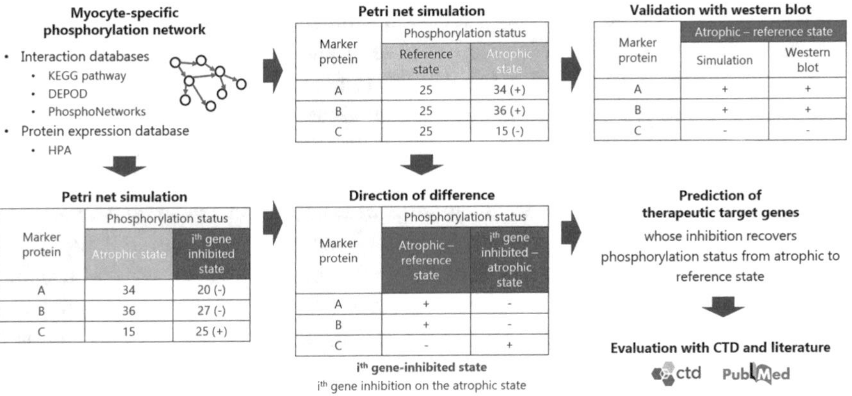

Fig. 8.1 Overview of predicting therapeutic target for muscle atrophy based on Petri net simulation of phosphorylation status

authors constructed a myocyte-specific network by adopting the same strategy used in [1]; trimming the backbone network by removing genes that are not expressed in the myocyte.

Then, the authors simulated the phosphorylation status of proteins in three types of states: normal (reference), atrophic state, and a specific-gene inhibited state. Petri net, a mathematical model for distributed systems, was used for simulations. It is advantageous for analyzing the dynamics of a large-scale network because it does not require detailed kinetic parameters, which are hardly available for a large-scale network. Petri net is a directed bipartite graph composed of two types of nodes: places and transitions. Places represent conditions or resources and transitions represent events. Places may hold tokens, which represent whether the condition of the corresponding place is met or not. If all the input places to a specific transition have a sufficient amount of tokens, then the transition can fire. The firing of a transition moves tokens from input places to output places.

For simulation, authors devised a Petri net model that is specifically suited to predict therapeutic targets based on the phosphorylation status of proteins. In the model, places represented proteins, and tokens represented phosphates so that the number of tokens in a place indicates phosphorylation status of the corresponding protein. Also, transitions represented phosphorylation or dephosphorylation reactions where input places correspond to kinases or phosphatases and output places correspond to their substrates. Unlike typical Petri net models where only one type of place and transition exists, authors classified places and transitions into two types, respectively. A place was classified based on whether the corresponding protein is activated by phosphorylation or dephosphorylation. If a protein is activated by phosphorylation, then the corresponding place enabled outgoing transition when it has enough amount of tokens. On the other hand, if a protein is activated by dephosphorylation, then the corresponding place enabled outgoing transition when it has few tokens. Transitions were classified into phosphorylation and dephosphorylation reactions. Phosphorylation transitions increased the number of tokens in an output place, whereas dephosphorylation transitions decreased the number of token in an output place. Unlike typical Petri net model, the number of tokens in input places remained unchanged since kinases and phosphatases are enzymes and phosphorylation status of themselves do not change during a reaction.

As the results of the simulation, authors obtained phosphorylation status of proteins at steady-state from normal (reference) state, atrophic state, and a specific-gene inhibited state. Then, authors compared the effect of muscle atrophy on reference state, and the effect of inhibition of a particular gene on the atrophic state. If inhibition of a gene increases phosphorylation of protein markers that are less phosphorylated in atrophic state and vice versa, the gene was suggested as a therapeutic target. They validated the results against literature and known muscle atrophy-related genes.

Interestingly, the authors found that the shortest path length from a gene to marker proteins was not indicative of therapeutic targets. This highlights the need for dynamic simulation of biological networks. Network dynamics can be analyzed by various tools such as ordinary differential equation (ODE), Petri net, and logic models. Among them, ODE is the most precise model that can predict the exact amount of modeled entities. However, ODE requires sufficient kinetic parameters, which often requires a vast amount of time-series data for their determination. This prohibits the application of ODEs to large-scale networks. For large-scale networks, Petri net or logic models can be used for simulating dynamics without kinetic parameters. Between them, Petri net was particularly suited for predicting therapeutic targets based on the phosphorylation status of proteins, because Petri net can adequately represent two types of nodes, activated by phosphorylation or dephosphorylation, and two types of edges, phosphorylation, and dephosphorylation. Even though unnecessariness of a large amount of data and kinetic parameters enabled Petri net and logic models to be applied to large-scale networks, it limits the level of detail the models can analyze. Petri net and logic models are usually limited to qualitative analysis rather than quantitative ones. For example, in [3], the authors focused on whether the phosphorylation of marker proteins increased or decreased, and did not pay attention to the degree of increase or decrease.

8.3 Prediction of Drug Repositioning

Finding new indications of already-used drugs can dramatically reduce cost and time because those drugs already satisfy numerous safety evaluations and skip the early stage of drug development. We call it "drug repositioning." To screen all existing drugs with all possible indications, computational methods are necessary. Especially, network-based approaches have been promising for drug repositioning as it can consider effect types and direction of drugs at the same time.

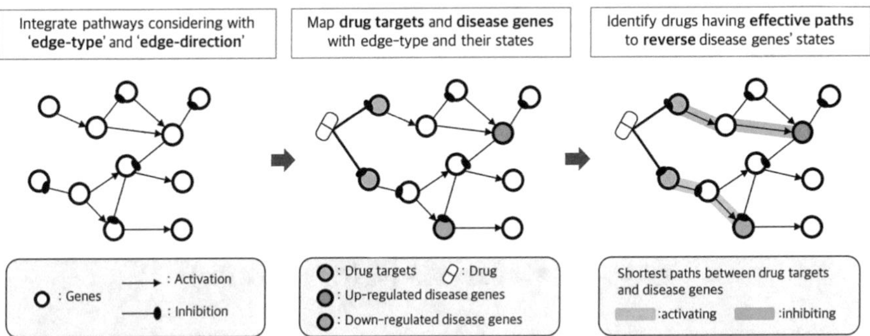

Fig. 8.2 Overview of predicting drug repositioning based on networks having effect types and effect direction

A work in [7], constructed drug–drug target interaction networks consist of effect types of interactions such as activation and inhibition with directions to find the new indication of drug (Fig. 8.2). Authors constructed a molecular network by extracting directed interactions between genes with detailed effect types (e.g. activation, inhibition) from [1] and assign different weights in their networks depending on the effect types of edges. Drugs are represented by corresponding target genes since the constructed network consists of genes. Their main hypothesis is that a drug having opposites effects on disease genes can alter states of disease genes and become effective. That is, if a drug elevates down-regulated disease genes or depresses up-regulated disease genes in the opposite way, it might be effective for the disease.

To measure how a drug can affect a certain disease, researchers often use a scoring function using the shortest path length between drug and disease. However, they needed a new scoring function between drug targets and disease genes because there can be contradictory effect types in a directed network. There are two kinds of reasons. One is that more than two shortest paths with different effect types can exist between drug targets and disease genes in a directed network. For example, TAK1 works with their downstream genes through two contradictory ways (both activation and inhibition exist). The other is that effect types can vary for each disease state. To solve the first problem, they tried to devise a new scoring function considering more than one shortest path and effect type.

$$d_c(r, g) = \frac{n_a + n_i}{n_a - n_i} |d(r, g)|$$

where n_a is the number of "activation-like shortest paths," n_b is the number of "inhibition-like shortest paths," and $|d(r, g)|$ is the absolute value of the shortest path length between drug targets and disease genes. In one of the shortest path, if an even number of inhibition edges exist, they treated as activation-like path and if an odd number, as an inhibition-like path. With this, they devised a new scoring function like below

$$S\,(\mathrm{R},\mathrm{G}) = \frac{1}{1+\left|\frac{d_c(r_i,g_j)}{\alpha}\right|^2}\cdot\frac{1}{n_g n_r}\sum_{i=1}^{n_r}\sum_{j=1}^{n_g}\mathrm{sgn}\left(r_i, g_j\right)$$

$$\mathrm{sgn}\left(r_i, g_j\right) = i\left(r_i\right)\cdot \mathrm{sign}\left(d_c\left(r_i, g_j\right)\right)\cdot s\left(g_j\right)$$

n_r is the number of drug targets of drug R, n_g is the number of disease genes of the disease G. $i(r_i)$ is the value representing the drug's effect and gets +1 for activation and −1 for inhibition. s(g_j) is the value representing the disease gene's state and gets +1 for the case of down-regulated and −1 for up-regulated. α is the parameter for making bell-shaped function. Their score increases when drug targets of a drug are close to disease genes and activate down-regulated genes or inhibit up-regulated genes in patients. To solve the second problem, they constructed a total of nine networks depending on diseases including various types of cancers and neurodegenerative diseases such as Parkinson's disease and schizophrenia based on experimental evidence.

Researchers can use their model to infer new drug indications and optimize the parameters of scoring function depending on their own purposes. However, this model has several drawbacks. First, we need a numerous number of effect types and effect direction information. If we do not have enough information for a certain disease, we cannot apply this model to predict drug repositioning. Secondly, effect types can sometimes become very ambiguous due to the absence of standards. Biological vocabularies such as "binding," "association," and "phosphorylation" cannot be clearly divided into activation or inhibition. This limitation can also be applied to effect direction. It is hard to come up with directions for some vocabularies.

8.4 Prediction of Drug Sensitivity

Cancer is highly heterogeneous not only between different cancer types but also between patients with the same cancer type [8]. There are two major reasons for this heterogeneity in cancer diseases. First, driver genomic variations are very different across cancer types [9]. Second, the different mutational landscape leads to a difference between patients with the same cancer type [10]. Following different mutation profiles, drug response in each person can also be different. Even in one patient, because of cancer evolving or resistance, drug sensitivity can change. For example, gefitinib suppresses RAS, NEK/ERK, and PI3K/AKT signaling pathways resulting in cell cycle arrest and apoptosis. However, some patients become resistant to the drug and the secondary RTK activates the aforementioned pathways again by

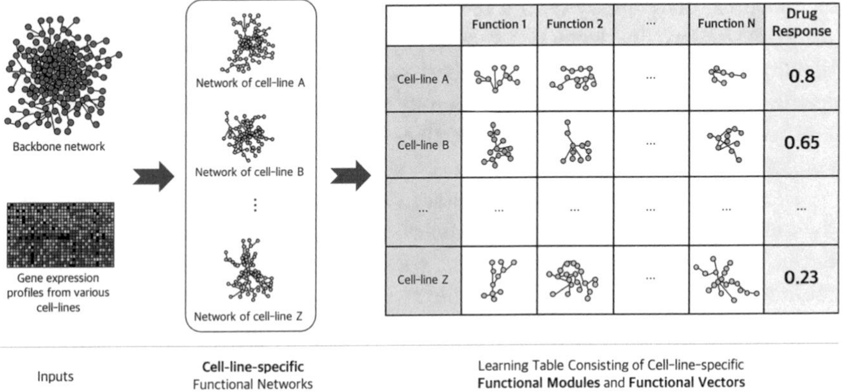

Fig. 8.3 Overview for predicting context-specific drug sensitivity

network rewiring. Therefore, it is very important to consider network rewiring or biological context to predict drug efficacy accurately.

A work in [11] developed a model that can predict drug sensitivity using cell line-specific functional modules (Fig. 8.3). Their main hypothesis is that all cancer types would be represented as different biological networks and show different functional modules. Therefore, the identification of different modules might help to predict precise drug sensitivity. To test their idea, they constructed a molecular network by extracting from [1]. They used protein–protein interactions and gene regulatory interactions to make a backbone network. However, it was not sufficient to show cell line-specific features so they collected gene expression data of nine different cancer types and corresponding normal cells to construct the cell line-specific network. Comparing cancer and matched normal state, they could find genes showing remarkable changes, thereby constructing nine cell line-specific networks.

The authors used a clustering algorithm to extract functional modules for each network. The clustering algorithm divides nodes of networks into a few modules depending on the network structure. Then they gave a score to each edge by calculating the Pearson Correlation Coefficient (PCC) between two nodes of gene expression profiles. Clustering can show how the elements of each module are similar to each other. However, it is hard to find the meaning of each group. Therefore, they devised a "Function vector" to identify real biological functions.

A function vector is a vector consisting of Gene Ontology (GO) terms enriched in genes of a functional module. A functional module consists of many genes so that they can perform enrichment analysis using those genes. After the enrichment test, they can get GO terms, especially biological process, which can represent functions of those genes. Using function vectors, they could match each functional module to real biological processes. If two functional modules share exactly the same biological processes, their function vectors might be the same and if two functional modules have similar biological processes, their function vectors are likely to be very similar. To compare all functional modules in a network with all function vectors, they used a measure named Jaccard index which can help to compare similarity between two vectors. Comparing all functional modules with all function vectors and getting the highest score per function vector, each cell line-specific network has scores across the function vectors. Using the matrix and drug sensitivity score (GI50) per each cell line, they made a regression model that can predict GI50 values.

They also analyzed relations between cell line-specific functions and drugs for another validation. In their predicted results, a drug, lapatinib is related to the immune system development and regulation of JAK-STAT cascade and cell proliferation. Lapatinib is known to block EGFR which is related to cell signaling pathway for cell growth so we can say their results are quite reliable. However, most biological processes in GO do not explain the direction of their phenomenon. For example, a function vector is revealed to be related to "RNA splicing" but we do not know whether this function activates RNA splicing or not. It is certain that their model helps to find cell line-specific drugs but to be more precise, we need more cell line-specific expression data and sophisticated GO term analysis.

8.5 Prediction of Drug–Drug Interaction

As bio-molecules have an interplay, a perturbation in a certain bio-molecule such as drug treatment or genetic mutation can influence neighboring molecules. Network analysis benefits from dealing with both close and distant effect of perturbation. One popular algorithm for analyzing signal propagation in networks is random walk (RW) algorithm that simulates the influence of nodes of interest on their neighbors based on network topology. Random walker starts from the nodes of interest (so-called seeds) and transits neighboring random nodes. At every step, each node has a probability that represents an influence from seed nodes. RW simulation ends once the probability for all nodes in the network is saturated. RW simulation with biological networks has been utilized, for example, for prioritizing novel disease-associated proteins or RNAs [12], predicting drug–target interactions [13], or revealing drug–disease associations [14].

In [15], the RW algorithm was applied to protein–protein interaction networks extracted from [1] to simulate drug effects in the molecular level and to predict drug–drug interactions (Fig. 8.4). As multiple prescription drugs (so-called polypharmacy) is increasingly common, adverse drug events are emerging due to unexpected interaction between drugs, resulting in the drug withdrawal. For drug–drug interaction analysis, authors developed a model that simulates influence on proteins of each drug and then calculates the interaction score of two drugs. To this end, a protein–protein interaction network is constructed, consisting of physical interactions between proteins. Random walker starts from target proteins of a drug. As periodic drug treatment continuously modulates target proteins, authors adopted the extended version, random walk with restart (RWR) [16] where random walker restarts from the seed nodes every step. Given r, the restarting probability of the

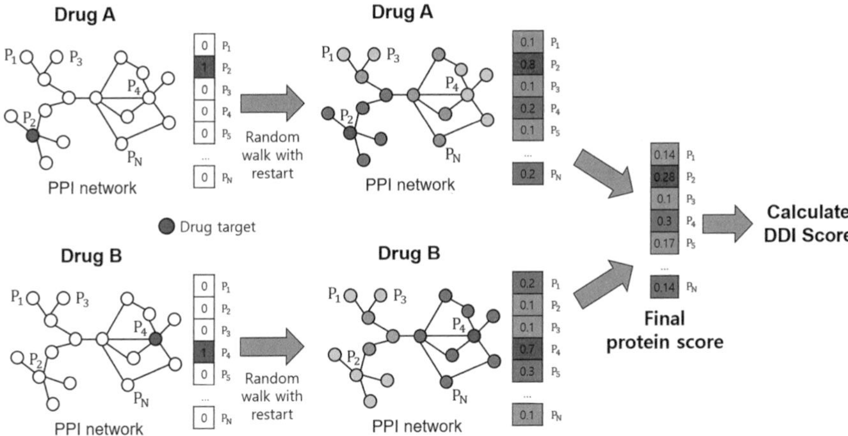

Fig. 8.4 Overview of predicting pharmacodynamic interaction between drugs using random walk with restart on protein–protein interaction network to predict

random walker at each time step, W, the normalized adjacency matrix of the PPI network, $p(t)$, the probability vector being each node at time step t, $p(0)$, initial probability vector, RWR algorithm is defined as:

$$p\,(t+1) = (1-r)\,W^T\,p(t) + r\,(p0)$$

Drug–drug interaction score is calculated using a defined scoring function as follows:

$$\mathrm{DDIScore}\left(\mathrm{Drug_A}, \mathrm{Drug_B}\right) = \sum_{i}^{N} \mathrm{ProteinScore}_i\left(\mathrm{Drug_A}, \mathrm{Drug_B}\right)$$

$$\mathrm{ProteinScore}_i\left(\mathrm{Drug_A}, \mathrm{Drug_B}\right) = \sqrt{V_i\left(\mathrm{Drug_A}\right) \times V_i\left(\mathrm{Drug_B}\right)}$$

where $V_i(\mathrm{Drug_A})$ represents the probability of protein i after RWR simulation of $\mathrm{Drug_A}$, and N represents the number of proteins in the network.

The performance of this model was evaluated using two different drug–drug interaction datasets which showed AUROC of 0.86 and 0.766, respectively. Two network-based models were selected for performance comparison. A model in [17] measures the distance between target proteins of two drugs, while the other model in [18] calculates overlapped proteins between two drugs' target-center systems that consist of target proteins and their first neighbors. For the two same datasets, [17] performed AUROC 0.807 and 0.696, whereas [18] performed AUROC 0.803 and 0.702. These results showed that considering both close and distant interferences of signaling propagation helps us to predict pharmacodynamic drug–drug interactions.

Biological application of random walks is advantageous to see both close and distance interference. In [15], random walk was suitably exploited to figure out pharmacodynamic interactions between drugs. However, they still have some limitations. Firstly, it is unable to elucidate the fact that molecular signals can be amplified as signals spread out. Second, as the influence on each protein from target proteins is uninterpreted for details such as whether it is an activating or inhibitory effect, the performance is limited in classifying the interaction types into antagonistic, synergistic, or additive effects. Enhanced versions of the RWR algorithm are expected to improve the simulation of signal propagation in biological networks in a more precise way.

8.6 Prediction of Hormone–Drug Interaction

One of the benefits of network analysis is the ability to elucidate detailed underlying mechanisms. A drug may have different responses to different patients who suffer from the same type of disease. Meanwhile, many efforts have been made to discover the factors that affect responses to drugs, and some studies have been addressed from a network point of view. In [19], the authors focused on hormones because the balance of hormones can vary a lot, depending on stress, diet, diseases, and drugs (Fig. 8.5). In particular, psychological stress induces the secretion of stress hormones such as epinephrine and cortisol, and many studies have shown that these hormones greatly reduce the therapeutic effect of drugs for the treatment of cancer. However, interactions between hormones and drugs remain poorly understood yet. Therefore, the authors aimed to create a model that can predict hormone–drug relationships in a systemic way and that can interpret underlying mechanisms of their crosstalks. To this end, the authors developed a model to analyze the mechanism of action (MOAs) of drugs and the influence of hormones on the MOAs. Here, hormone–drug relationships are examined in a disease-specific way because the same drug may have different drug MOAs depending on the disease to be treated. First, a molecular network was constructed by extracting directed

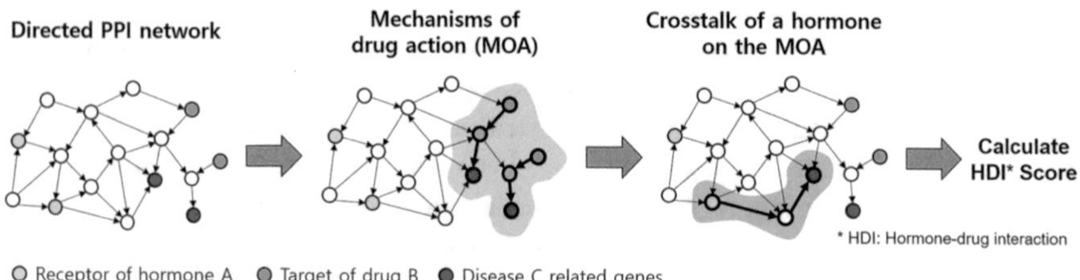

Fig. 8.5 Overview of predicting the influence of a hormone on the response of a drug for the treatment of a disease

and physical interactions between proteins from [1]. Hormone receptors, drug targets, and disease-related genes are mapped on the network in order to represent a hormone, a drug, and a disease, respectively. To predict the MOAs of the drug, the model identifies the shortest paths from each of the target proteins to the closest disease gene, and the paths were defined as the drug effect paths (DEPs). To analyze the effect of the hormone on the drug MOAs, the shortest paths from receptors of the hormone to the closest protein on the DEPs were identified, and the paths were defined as the hormone effect paths (HEPs). The scoring function for calculating the potential influence score of the hormone on the response of the drug was defined based on the assumption that as hormones modulate receptors closer to drug MOAs are more likely to affect the drug response. Given where $\boldsymbol{R}(\boldsymbol{h})$, the set of hormone receptors, $\boldsymbol{M}(\boldsymbol{d})$, the set of molecules of DEPs, $\boldsymbol{n}(\boldsymbol{S})$, the number of distinct start nodes, $\boldsymbol{n}(\boldsymbol{E})$, the number of distinct end nodes, $\boldsymbol{d}(\boldsymbol{r}, \boldsymbol{m})$, the shortest path length from receptor $\boldsymbol{r}$ to molecule $\boldsymbol{m}$, and $\boldsymbol{\alpha}$, a decay constant, the potential impact $\boldsymbol{i}$ for a hormone $\boldsymbol{h}$ and a drug $\boldsymbol{d}$ is defined as:

$$\boldsymbol{i}\,(\boldsymbol{h}, \boldsymbol{d}) = \boldsymbol{\alpha}^{-\min_{\boldsymbol{r}\in \boldsymbol{R}, \boldsymbol{m}\in \boldsymbol{M}(\boldsymbol{d})} \boldsymbol{d}(\boldsymbol{r}, \boldsymbol{m})} \times \boldsymbol{n}\,(\boldsymbol{S}) \times \boldsymbol{n}\,(\boldsymbol{E})$$

This model was tested for 20 diseases including five cancers and 15 non-cancer diseases and performed AUROC 0.89 on average at $\alpha = 8$. Since this model is the first systems approach for predicting interference of hormones on the efficacy of drugs, two closely related methods were selected for performance comparison. A model introduced in Sect. 8.5 [15] was developed to predict drug–drug interactions and was based on RWR simulation. The other model in [20] was created to predict drug–target interactions, based on known drug-target information. For 13 diseases applicable for the two comparable models, [15, 19, 20] performed averagely AUROC 0.88, 0.84, and 0.57, respectively. Also, the authors demonstrated that known hormone-drug pairs whose hormones have impacts on the responses of the drugs have shorter HEP lengths compared to unlabeled hormone-drug pairs. These results demonstrated that the shortest path-based strategy was helpful for analyzing crosstalks between hormones and drugs.

As shown in the above study, network analysis is useful in predicting the functional interaction between two elements and interpreting its detailed mechanism. The model developed in [19] can be applied to other relationships similar to hormone–drug relationships. However, this model has some limitations. First, there may be exceptional cases where the shortest pathway is not suitable for the mechanism analysis, unlike DEPs and HEPs. Second, this model can calculate interaction score, but it does not tell us detailed interaction types (e.g. positive effect or negative effect). If omics data is additionally integrated or path analysis algorithms are improved, we expect this model to be used for analyzing diverse relationships including hormone–drug relationships.

8.7 Prediction of Therapeutic Effects of Natural Products

Biological networks can be used for predicting the therapeutic effects of not only drugs but also natural products. Natural products are attractive chemicals for drug development as they are proven to be bioactive in some organisms [21]. Since there are various kinds of natural products, computational methods are required to analyze their therapeutic effect. In [22], the authors analyzed the therapeutic effect of natural products under the hypothesis that natural products will have a similar effect on human metabolite if they are similar to human metabolite (Fig. 8.6). Authors measured the similarity between a natural product and a human metabolite in terms of three criteria: structure, target, and phenotype.

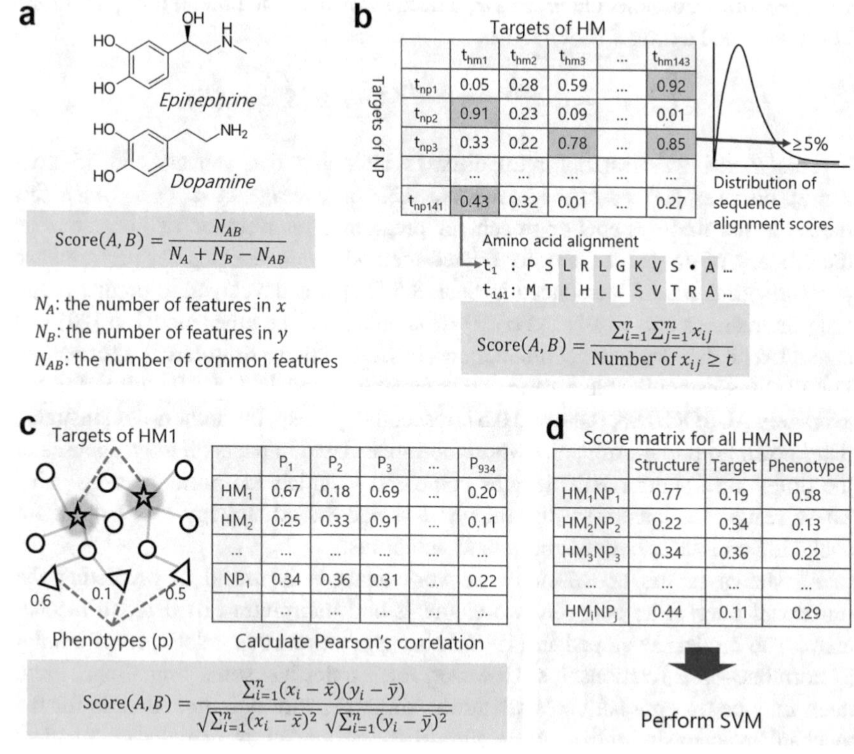

Fig. 8.6 Overview of the three similarity measures used to evaluate the similarity between a natural product and a human metabolite. (**a**) Measure structural similarity between a human metabolite and a natural product using Tanimoto coefficient. (**b**) Measure amino acid sequence similarity of two target proteins. (**c**) Measure phenotype similarity between a human metabolite and a natural product using Pearson's correlation score where vectors of all phenotype scores are generated by the random walk restart algorithm on the CODA network. (**d**) Calculate final score matrix for all human metabolite and natural product pairs and then train SVM model

Structure similarity of compounds is one of the most popular features that are used for analyzing the therapeutic effects of compounds [23]. The underlying hypothesis is that structurally similar compounds are more likely to bind to the same target, which leads to a similar therapeutic effect. Based on the hypothesis, the authors calculated structure similarity for each pair of a natural product and a human metabolite. Structure similarity was calculated by using Tanimoto coefficient which measures how many fingerprints are overlapped between two chemicals.

However, high structural similarity between two compounds does not necessarily mean their targets are also similar. To supplement structure similarity, the authors used target similarity as well. For each pair of a natural product and a human metabolite, authors measured how much similar amino acid sequences of their targets were. For each pair of a target of a natural product and a target of a human metabolite, amino acid sequence similarity was calculated by using a Smith–Waterman sequence alignment score. Then, the top 5% sequence similarity scores among all pairwise target sequence similarity scores were selected and averaged to obtain the target similarity score.

However, our knowledge of targets of natural products and human metabolites is incomplete. To compensate this, authors also evaluated phenotype similarity between a natural product and a human metabolite. To obtain the similarity, the authors firstly constructed a network composed of various biological entities such as genes, metabolites, cellular functions, and phenotypes. Then, the authors applied the RWR algorithm to the biological network by using each natural product and human metabolites as a seed node. After the simulation, authors obtained a score for each node in the network which quantifies the effect of a natural product or a human metabolite on them. This means that we can evaluate the effect of a natural product or a human metabolite on various biological entities including cellular functions or phenotypes by one simulation. Among various types of nodes, authors focused on phenotype nodes. They calculated Pearson's correlations between the scores of phenotypic nodes for measuring phenotype similarity between a natural product and a human metabolite. If a natural product and a human metabolite locate in close vicinity of within the biological network, they are more likely to have high phenotype similarity score. Finally, the authors used three similarity scores—structure, target, and phenotype similarity scores—to determine whether a natural product and a human metabolite are similar or not. Then, the therapeutic effect of a natural product was predicted as the function of similar human metabolite.

Since there is no gold standard data stating which natural product is similar to which human metabolite, authors used drugs instead of natural products that resemble human metabolites as a silver standard data for model evaluation. Authors showed that all three similarity measures were capable of identifying similar drug and metabolite pair, including phenotype similarity calculated on the biological network. Among them, structure similarity has the highest predictive power. However, combining three similarity measures increased the performance, indicating that three similarity measures were complementary to each other.

In [22], the authors suggested that the phenotype similarity calculated by an RWR algorithm on the biological network can be helpful for predicting the therapeutic effect of a natural product. However, they only cared about how similar the effect of a natural product on phenotypes is to that of a human metabolite, while not considering phenotypes that were predicted to be highly affected by a natural product. Incorporating cellular functions or phenotypes that were highly affected by a natural product by the RWR algorithm might help us to predict the therapeutic effect of a natural product.

8.8 Conclusion

In this chapter, we explored the CODA network and six computational models that applied network analysis techniques on the network for drug development. Complex biological phenomena can be interpretable when considering not only individual effects of biological entities but also inter-connections between them. The great advantage of biological networks is the ability for comprehensive and systematic analysis of the complex biological events. Therefore, biological network-based approaches are attracting increasing interests.

In the six studies above, authors constructed networks consisting of biological events extracted from the CODA network and conducted network analysis by employing the basic network principles and/or novel creative algorithms. For networks, cell/disease-specific network, directed signaling network, undirected protein–protein interaction network, or multi-layer networks were utilized. For network analysis, Petri net simulation, random walk simulation, pathway identification, functional module detection, and pathway crosstalk analysis were applied. The six approaches still have rooms to be improved for better performance or interpretation, but they outperformed other comparable methods. The overall discussion in this chapter leaves an important message that studies of different purposes require different strategies on networks and analysis algorithms.

References

1. Yu, H., et al.: CODA: integrating multi-level context-oriented directed associations for analysis of drug effects. Sci. Rep. **7**, 1–12 (2017)
2. Kwon, M., et al.: CODA-ML: context-specific biological knowledge representation for systemic physiology analysis. BMC Bioinform. **20**(10), 248 (2019)
3. Jung, J., et al.: Petri net-based prediction of therapeutic targets that recover abnormally phosphorylated proteins in muscle atrophy. BMC Syst. Biol. **12**(1), 26 (2018)
4. Kanehisa, M., Goto, S.: KEGG: Kyoto encyclopedia of genes and genomes. Nucleic Acids Res. **28**, 27–30 (2000)
5. Hu, J., et al.: PhosphoNetworks: a database for human phosphorylation networks. Bioinformatics. **30**, 141–142 (2014)

6. Duan, G., Li, X., Köhn, M.: The human DEPhOsphorylation database DEPOD: a 2015 update. Nucleic Acids Res. **43**, D531–D535 (2015)
7. Yu, H., et al.: Prediction of drugs having opposite effects on disease genes in a directed network. BMC Syst. Biol. **10**(1), S2 (2016)
8. Fisher, R., Pusztai, L., Swanton, C.: Cancer heterogeneity: implications for targeted therapeutics. Br. J. Cancer. **108**(3), 479 (2013)
9. Yi, S., et al.: Functional variomics and network perturbation: connecting genotype to phenotype in cancer. Nat. Rev. Genet. **18**(7), 395 (2017)
10. Wedge, D.C., et al.: Sequencing of prostate cancers identifies new cancer genes, routes of progression and drug targets. Nat. Genet. **50**(5), 682 (2018)
11. Hwang, W., et al.: Context-specific functional module based drug efficacy prediction. BMC Bioinform. **17**(6), 275 (2016)
12. Liu, Y., et al.: Inferring microRNA-disease associations by random walk on a heterogeneous network with multiple data sources. IEEE/ACM Trans. Comput. Biol. Bioinform. **14**(4), 905–915 (2016)
13. Chen, X., Liu, M.-X., Yan, G.-Y.: Drug–target interaction prediction by random walk on the heterogeneous network. Mol. BioSyst. **8**(7), 1970–1978 (2012)
14. Liu, H., et al.: Inferring new indications for approved drugs via random walk on drug-disease heterogenous networks. BMC Bioinform. **17**(17), 539 (2016)
15. Park, K., et al.: Predicting pharmacodynamic drug-drug interactions through signaling propagation interference on protein-protein interaction networks. PLoS One. **10**(10), e0140816 (2015)
16. Köhler, S., et al.: Walking the interactome for prioritization of candidate disease genes. Am. J. Hum. Genet. **82**(4), 949–958 (2008)
17. Gottlieb, A., et al.: INDI: a computational framework for inferring drug interactions and their associated recommendations. Mol. Syst. Biol. **8**(1), 592 (2012)
18. Huang, J., et al.: Systematic prediction of pharmacodynamic drug-drug interactions through protein-protein-interaction network. PLoS Comput. Biol. **9**(3), e1002998 (2013)
19. Kwon, M., et al.: HIDEEP: a systems approach to predict hormone impacts on drug efficacy based on effect paths. Sci. Rep. **7**(1), 16600 (2017)
20. Cheng, F., et al.: Prediction of drug-target interactions and drug repositioning via network-based inference. PLoS Comput. Biol. **8**(5), e1002503 (2012)
21. Harvey, A.L.: Natural products in drug discovery. Drug Discov. Today. **13**, 894–901 (2008)
22. Noh, K., Yoo, S., Lee, D.: A systematic approach to identify therapeutic effects of natural products based on human metabolite information. BMC Bioinform. **19**(8), 205 (2018)
23. Martin, Y.C., Kofron, J.L., Traphagen, L.M.: Do structurally similar molecules have similar biological activity? J. Med. Chem. **45**, 4350–4358 (2002)

Chapter 9
Network Propagation for the Analysis of Multi-omics Data

Minwoo Pak, Dabin Jeong, Ji Hwan Moon, Hongryul Ann, Benjamin Hur, Sangseon Lee, and Sun Kim

Abstract Network propagation has been used as the state-of-the-art network analysis method which can overcome the limitations of the existing methods. The area where network propagation truly shines is the application to multi-omics data. It integrates interactome data with other omics data to be utilized for further analyses. In multi-omics data, there are multiple levels of complexity for which network propagation techniques can be used effectively. Here, we categorized recent network propagation research works in bioinformatics into three levels: DNA, RNA, and pathway levels. At the DNA level, network propagation is applicable for genome data analysis. In particular, for the analysis of genome-wide mutation profiles in cancer, network propagation has been successfully used to detect subnetworks, driver mutation profiles, and cancer subtypes. At the RNA-level analysis of transcriptome data, gene expression information can be propagated to obtain valuable condition-specific information such as the ranking of how genes react to certain environments. Such derived information can be utilized in differentially expressed gene detection, gene prioritization, and hypothesis testing. Transcriptome data can be analyzed at the biological pathway level by utilizing network propagation techniques. With graph convolutional network techniques, network propagation can be exploited for a disease subtype classification problem where pathway information is used to characterize biological mechanisms of a disease.

Keywords Network propagation · Multi-omics · Genome · Transcriptome · Pathway

M. Pak · J. H. Moon · H. Ann · B. Hur · S. Lee · S. Kim (✉)
Department of Computer Science and Engineering, Seoul National University, Seoul, South Korea
e-mail: mwpak@snu.ac.kr; sunkim.bioinfo@snu.ac.kr

D. Jeong
Interdisciplinary Program in Bioinformatics, Seoul National University, Seoul, South Korea

© Springer Nature Switzerland AG 2021
B.-J. Yoon, X. Qian (eds.), *Recent Advances in Biological Network Analysis*,
https://doi.org/10.1007/978-3-030-57173-3_9

9.1 What Is Network Propagation

For the investigation on relationships among multiple entities in a complex system or group, it is a common practice to conduct network analysis. One of the main advantages of network analysis is that it is easy to analyze how entities form relationships. Even with a large number of entities, it is easy to see which entities actively interact with others and which are relatively isolated. We can also explicitly use a graph-based clustering or community detection algorithm to visualize what patterns or groups are formed. In other words, hidden patterns, which are not distinct in raw data, can be captured through network analysis. Due to such effectiveness in detecting patterns and capturing complex relations, development of network analysis methods has been an active research topic in various fields. The field of bioinformatics is no stranger to network analysis. Various researches have been conducted to characterize interactions between biological entities that collectively represent a phenomenon or phenotype such as disease or drug response.

However, most network analysis methods assume that network is static. This is a significant hurdle when investigating the dynamics of how entities interact with one another over time.

In addition, early network analyses in bioinformatics were based on the principle of *guilt by association* [1]. Guilt by association principle assumes that biological components—such as genes or proteins—are closely related in terms of their biological role if they are direct neighbors to one another in the network. However, this is not always true. For instance, proteins that are direct neighbors in the protein–protein interaction (PPI) network do not necessarily participate in similar biological activities or share similar characteristics. Namely, network analysis approaches that rely on such assumption would introduce high false positives. This is where network propagation comes in.

Network propagation is the state-of-the-art network analysis method that has been developed and used recently. The characteristics that differentiate network propagation from existing methods include its ability in handling temporal information and topology of the whole network. That is, network propagation considers node relationships of the whole network, not just the neighbors within a certain distance. Due to such advantages, network propagation techniques are being utilized across many different fields with different settings and also different names. For example, random walks on graphs, information diffusion, electrical resistance, and graph kernels are different forms of graph analysis techniques that all use network propagation as the core technique. In biology, main applications of network propagation include gene function prediction, disease subtype classification, drug target prediction, disease module discovery, and gene prioritization as surveyed in a recent review paper [1].

9.1.1 How Network Propagation Works

The process of network propagation is often described in analogy with fluid or heat diffusion. Let us assume that we have a network with multiple nodes and edges, where nodes represent proteins and edges represent interactions between proteins. Based on prior knowledge, suppose that some nodes are labeled with a property that is likely to be shared within its local neighborhood. The labels of the nodes can be thought of as *fluid* of some fixed amount. Namely, each node which is labeled with the property of interest is given a bucket of fluid that can be diffused throughout the network via edges between nodes. With this setting, the network is ready to begin the propagation. At each propagation step, fluids at the labeled nodes flow to the direct neighboring nodes either by a fixed amount or by amounts that are proportional to edge weights. This flow of fluids toward direct neighbors in each time step is repeated for a certain number of steps such that most of the fluids are still near the original labeled nodes. This is to keep the result meaningful because once the fluid is diffused out through the whole network, the result would not be informative. By examining the amounts of the fluid of nodes at each intermediary time step, one can obtain valuable information on how the influence is transmitted hence better understand how the process of interest is carried out and which entities are involved in it [1, 2].

! Attention

To elaborate on the process, we discuss the mathematical aspects of network propagation. In order to perform network propagation, one needs two essential components: a network and seeds. A network is represented as a weighted adjacency matrix W, where the value in an entry within W represents the weight of the corresponding edge. A set of seed nodes that correspond to entities in a certain domain, typically genes in bioinformatics, are represented as a vector $\mathbf{p_0}$. The initial values of $\mathbf{p_0}$ represent the measure of confidence or significance of the seed nodes according to prior knowledge. The values of the vector $\mathbf{p_0}$ determine how the amount of fluids is initially distributed in the network. Note that $\mathbf{p_0}$ can also be discrete such that seed nodes have the value of 1 and the remaining nodes have the value of 0. With the weighted adjacency matrix W and the vector $\mathbf{p_0}$, the scores of the seed nodes are propagated through the directly interacting genes as follows:

$$\mathbf{p_t} = W\mathbf{p_{t-1}}. \tag{9.1}$$

At every time step, the weighted adjacency matrix W and the node score vector $\mathbf{p_{t-1}}$ are multiplied to obtain the new node score vector at the next time step t. Conceptually, a node score at time step t is calculated through weighing the node scores of its neighboring nodes by the corresponding edge weights and updating the node score proportional to the sum of the resulting values. In other words, each node score is updated proportional to the weights of the edges to the neighboring nodes as well as their node scores. This is analogous to how the fluids flow to the direct neighbors according to the edge weights. As a result, genes that are reached by many seed genes receive high node scores. On the other hand, hub genes that have many outgoing edges are deprecated since hub genes can reach many nodes including irrelevant genes. Note again that the network propagation process should not continue for too many steps since the fluid can be diluted throughout the network and the node scores would be smoothed out.

To address this dilution problem, the random walk with restart (RWR) algorithm is infused to the propagation process. This algorithm is also called Markov Random Walk (MRW) since the randomness of diffusion process is stochastically modeled with the edge weight W.

$$\mathbf{p_t} = \alpha \mathbf{p_0} + (1 - \alpha) W \mathbf{p_{t-1}} \tag{9.2}$$

By adding the new term α and the initial state vector $\mathbf{p_0}$, we can weigh the trade-off between the prior information and the dilution of information. Restarting probability α decides how much information from the original profile $\mathbf{p_0}$ is retained. For example, if α is 0.7, 70% of information from the original profile is retained and 30% of information is propagated from the neighboring nodes and added to the current status, $\mathbf{p_t}$, of the entities. Therefore, a whole new profile, $\mathbf{p_t}$, is updated from $\mathbf{p_{t-1}}$, considering the network topology. Iteratively, this information is transmitted along the direction of interactions until convergence. The profile of nodes after convergence represents the order of priority, where the entity with the highest score has the highest priority.

9.1.2 Properties of Network Propagation

To illustrate the power of network propagation in action, let us take the gene prioritization problem as an example. Given a network and a list of genes that are already known to be associated with a certain disease, the goal of gene prioritization is to find the candidate genes that are previously not known to be associated with the disease and prioritize them in a ranked list.

One approach without using the network propagation would be to look at the direct neighbors of the known genes. However, because of the aforementioned *guilt by association* principle, this approach would introduce many false positives. Another approach would be to search through not only direct neighbors but also indirect neighbors by considering genes that are reachable by a fixed short distance. Unfortunately, this approach would be encountered with another well-known problem in biological networks called the *small world* property that indicates that almost all genes in most biological networks are reachable by one another in a small number of steps. In this case, we would be confronted with many false negative results. Furthermore, both approaches take only parts of the network into account, not the whole structure.

Network propagation provides solutions to the challenges of these methods. Unlike other approaches including the shortest path-based approach, network propagation considers the topology of the whole network. Network propagation diffuses the fluid from the seed nodes through all connected edges. Depending on how many steps the fluid is propagated, it would reach different parts of the network with different amounts. The amounts of the fluid that the unlabeled nodes possess at the end of the propagation process can be used as degrees of significance related to the property of interest. By taking account of all possible paths between genes to give nodes different degrees of importance, network propagation provides a much more refined result with less false positives and false negatives. In other words, genes that are connected with the seed nodes by a small number of paths are down-weighted, while genes with many possible paths from the seed genes are promoted. Namely, genes that would have been missed by the shortest path-based approaches despite its high connectivity receive higher node scores.

In addition, network propagation can also adjust for local connectivity while avoiding giving too much credit to hub nodes. Nodes that receive multiple paths from the labeled nodes would get more fluid. On the contrary, nodes that have many outgoing edges would lose their fluid quickly. These properties are what differentiate network propagation from the existing network analysis methods. In biology, these properties are especially useful not only for gene prioritization but also for many other various applications.

9.2 Genome Data Analysis at DNA Level

Genomics is an integrative study on a genome [3]. Genome, the code for all forms of life on earth, has been a key to decode how organisms work. As its replication process is deliberately regulated and repaired, these codes are assumed to be passed down to the next generation without any changes. Therefore, genetic diseases (e.g., cancer) are considered to be connected with the mutation on the genome. In other words, how mutations on genome are related to the phenotype is a clue to determine the cause of disease. Recently, rather than focusing on the effects of a mutation on each individual gene, studies that integrate the effects of mutations with a network have emerged. In this section, we will introduce two recent studies that used network propagation techniques successfully for the genome analysis: one for determining rare mutations in cancer [4] and another for classifying cancer subtypes [5].

9.3 Detecting Rare Somatic Mutations Using Network Propagation

Cancer is a disease of genetic mutation. Since cancer cells are heterogeneous, defining the characteristics of cancers is a complicated task. One of the main challenges confronted by the existing methods such as single-gene tests and network approaches is the *long-tail phenomenon*. The *long-tail phenomenon* characterizes the tendency of cancer-related genes that most of them are mutated at a low frequency rate [6]. Because of this phenomenon, it is hard to distinguish between cancer-related driver mutations and passenger mutations. In addition, it is also challenging to incorporate the topology of the whole network of interactions between genes in the analysis. In this section, we introduce a study called HotNet2 [4] that provides a solution to these challenges using network propagation. HotNet2 detects mutated subnetworks that contain cancer-related genes with not only high but also rare mutation frequencies while taking the topology of the whole interaction network into account. Namely, the algorithm finds the subnetworks with statistically significant cancer-related mutations in a broad-ranged datasets such as pan-cancer data.

HotNet2 algorithm works as follows (Fig. 9.1). First the seed information, which is propagated through the network, is the genomic variation information (e.g., mutation frequencies of genes) derived from somatic mutation profiles of single-nucleotide variants (SNVs) and copy number aberrations (CNAs). A special point about this algorithm is that it uses directed networks for network propagation rather than undirected networks. In an undirected network, an edge does not capture the information of direction between two adjacent nodes—for example, A and B. Hence, the edge weight from node A to node B is equal to the edge weight from B to A, which makes the influence between interacting nodes symmetric. However, in a directed network, two interacting nodes reciprocally give and take asymmetric influence. Therefore, nodes connected with mutual interactions are likely to be assigned with higher scores when propagated. Here, the weights of edges are defined proportional to the inverse of outdegree of node, in other words, genes that have a few interactions diffuse less heat to their neighbors and vice versa.

Extending the existing network propagation model—HotNet [7, 8], HotNet2 adopted insulated diffusion algorithm, which is basically random walk with restart (RWR) algorithm. The property assigned to the node is diffused to the neighbors iteratively, throughout the whole network. Since the prior information of seeds is considered, the diffusion process effectively captures the local topology of interacting nodes around seeds. After the propagation terminates, subnetworks are detected by removing the edges, which have the amount of head propagated below the threshold δ. Then, statistical tests—NMC [9] and iPAC algorithm [10]—filter out non-significant subnetworks.

Applied to the TCGA pan-cancer data of somatic mutation profiles, HotNet2 detects biological pathways, complexes, and genes related to the characteristics of cancer. Across multiple tumor types, significantly mutated subnetworks that

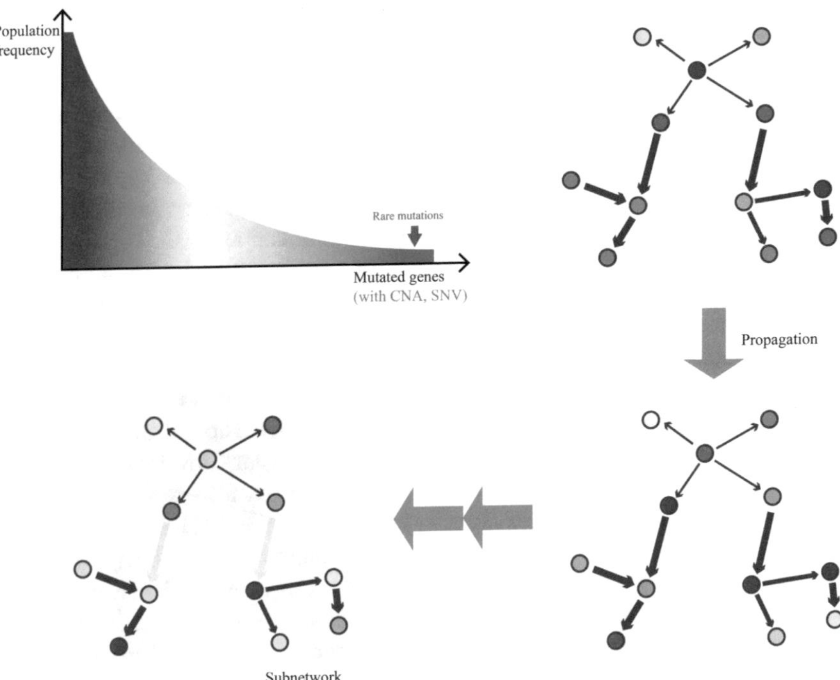

Fig. 9.1 Pan-cancer analysis with HotNet2. SNP and CNV data is utilized as seed for network propagation; population frequency of mutated genes are the property of nodes to be propagated. The process of pan-cancer analysis to identify significantly mutated subnetworks using HotNet2 on a broad-ranged TCGA dataset [4]

encompass important cancer pathways and complexes are detected by HotNet2. For example, reported cancer pathways and complexes (e.g., receptor tyrosine kinase signaling pathway, SWI/SNF complexes) and potentially important cancer pathways (e.g., major histocompatibility complex 1) are detected as subnetworks. These subnetworks include genes with low somatic mutational loads, which are interacting with well-known cancer-related proteins. This result gives additional evidence to the role of rarely mutated genes in cancer.

9.3.1 *Disease Subtype Classification Using Supervised Network Propagation*

Note again that tumor is a disease caused by the accumulation of genetic mutations. Up to 90% of mutations in cancer are somatic mutations according to The Cancer Gene Census [11]. In the development process of tumor, somatic mutations in genes turn them into malfunctioning status. How mutations are loaded in the genes in a tumor? Interestingly, every time the mutation occurs, it tends to be located at different genes. Only few genes, therefore, are commonly mutated and most mutations are scattered in the genome [12]. In other words, even mutation profiles of the same tumor type are diverse so that it is hard to understand how somatic mutations affect tumor. One of the prominent approaches is to classify tumors into subtypes based upon somatic mutation profiles: *tumor stratification*, the method that groups patients into hierarchical subtype based on molecular profile. Recently, network propagation has been applied to tumor stratification [5, 13, 14]. In network propagation, the mutation profile of a gene is iteratively diffused to the neighboring genes in the network until convergence. After convergence, all the effects of neighbors are summed up to reconstruct the mutation profile. The properties of each gene after convergence reflect the original properties of a gene as well as the properties of the network topology. Therefore, the reconstructed mutation profiles are less heterogeneous than the previous ones due to the incorporation of the connectivity of the network, which makes subtype classification easier. However, there is still niche left. The problem is that network propagation on a conventional network (e.g., PPI) cannot detect cancer-specific or tissue-specific subnetworks since the network stays stationary during propagation. Although certain pathways—survival or proliferation—are highly activated in cancer cells, a general network cannot reflect the characteristics of cancer. To detect domain-specific subnetworks, Wei Zhang et al. [5] proposed a novel method called *supervised network propagation*.

The special properties of supervised network propagation are that the network, where mutation profiles are propagated, is updated to represent the characteristics of subclass as the iteration proceeds. That is, an edge of the network is adjusted so that each subnetwork detected from the adjusted network represents each subtype of tumor. How the edges are adjusted? At every iteration, the edge weight is accommodated to minimize the loss function. Wei Zhang et al. [5] defined novel loss function that represents the difference between the real mutation profile and the propagated mutation profile of each subclass. Here is a clue to why this method is called *supervised* network propagation. The original mutation profile is needed to calculate the loss; hence, it is called supervised network propagation unlike the typical network propagation.

9.3.1.1 How Supervised Network Propagation Works

The molecular interaction network, where the mutation profile is propagated, is composed of nodes and edges. The edge between the entities is represented with multiple features—interaction type, source database, cancer-related pathways, etc. The weighted sum of features compresses the feature vector for each edge into a single value so that the network can be represented as a transition matrix with scalar elements. Another key player in network propagation is the seed value to be propagated in the network. Here, a mutation profile for each tumor is exploited as a seed. For each tumor, whether the mutation occurs or not is marked as a binary digit for every gene.

Using this transition matrix, a seed value is propagated along the network as in a general network propagation. The propagated mutation profile is compared with the original mutation profile to calculate the loss $J(\boldsymbol{\omega})$. The purpose here is to find a feature weight vector $\boldsymbol{\omega}$ that minimizes the loss function $J(\boldsymbol{\omega})$.

$$\min_{\boldsymbol{\omega}} J(\boldsymbol{\omega}) = \lambda||\boldsymbol{\omega}||_1 + \sum_{u=1}^{m} \frac{1}{1 + exp(-\beta D_u)} \tag{9.3}$$

The first term in the right-handed side is a regularization term for feature weight $\boldsymbol{\omega}$ with parameter λ. The second term denotes for the sum of normalized D_u for each subtype, where m is the number of tumor samples. Since the term D_u represents the difference between the original profile and the propagated profile for certain sample u, minimizing this term leads to proper classification.

Here is detail on how D_u is defined.

$$D_u = ||\mathbf{p}_u - \mathbf{c}_a||_2^2 - min_{b \neq a}||\mathbf{p}_u - \mathbf{c}_b||_2^2, \tag{9.4}$$

$$\mathbf{c}_a = \frac{1}{m_a - 1} \sum_{v \in a, v \neq u} \mathbf{p}_v. \tag{9.5}$$

Here, c_a stands for the centroid of subtype a, which is derived from the original mutation profiles before propagation. When the mutation profile is propagated, the reconstructed profile p_u is obtained. It is the representative of each subtype from the original mutation profile. The first term of D_u is the distance between the propagated vector p_u and the designated subtype a, while the second one is the distance between p_u and the centroid vector of second best subclass. To minimize D_u, the distance from the subtype which the sample u belongs to should be maximized and the distance from the second best subtype should be maximized; therefore, the subclass is distinctly classified.

If the convergence condition is not met, ω is updated by gradient descent. Then, the whole process is conducted iteratively until convergence. After the training is done, when the new sample z is given, the predicted subtype s is assigned as follows:

$$s = argmin_{s \in A}||\mathbf{p}_z - \mathbf{c}_s||_2^2. \tag{9.6}$$

Among all subtypes A, the distance between a propagated profile and the centroid vector of subtype s is the minimum, the subtype of sample z is assigned as s.

9.3.1.2 Where the Supervised Network Propagation Is Applied

As mentioned before, supervised network propagation can be used to reduce heterogeneity of cancer mutation profiles, i.e., in each of cancer subtypes. Glioblastoma is a malignant brain tumor, which can be classified into five subtypes based on gene expression and DNA methylation: Glioma-CpG Island Methylator Phenotype (G-CIMP), Proneural, Neural, Classical, and Mesenchymal [15, 16] . Using supervised network propagation, prediction accuracy was up to 64%, and precision and recall were high, overall. More importantly, a subnetwork detected by supervised network shows good interpretation on which biological networks are activated or not in tumor cancer. For instance, a classical subtype is known for single-nucleotide polymorphism (SNP) or copy number variation (CNV) on epidermal growth factor receptor (EGFR) gene. In the adjusted network derived from supervised network propagation, EGFR-LAMA1-PIK3CA-LRP2 subnetwork is heavily weighted. That is, EGFR-related interactions are detected as subnetwork through the supervised network propagation in a subtype where the EGFR is highly mutated. Similarly, breast cancer subtypes—i.e., Basal, Her2, Luminal A, and Luminal B [17]—are properly discernible, and subnetworks highlighted from the supervised network propagation capture the characteristics of the biological pathway of breast cancer (Fig. 9.2).

9.4 Transcriptome Data Analysis at RNA Level

Transcriptome data can be integrated with interactome data to be analyzed in order to gain condition-specific information related to various biological phenomena. Network propagation is a very useful tool in this area. An example would be to rank differentially expressed genes in various experiments while considering the information regarding the interactions among genes and the topology of the given network. In addition, gene expression levels can be propagated to obtain the ranking of how well genes or proteins of interest explain variations in the expression levels of other genes under certain conditions. The power of network propagation has been further exploited in many transcriptome data analyses. In this section, we discuss some of the recent studies that use the network propagation technique successfully for the analysis of transcriptome data.

9.4.1 Gene Prioritization

Gene prioritization is to rank genes in association with various phenotypes such as disease [18, 19] or drug target [20], or other interesting phenotype [21]. To rank genes, it is necessary to consider complex relationship among genes, which requires guidance or prior knowledge. Therefore, gene interaction information such as protein–protein interaction (PPI), transcription factor (TF)–target gene (TG) regulation, miRNA-TG regulation, and co-expression, is additionally used as biological prior knowledge in the form of networks to improve the accuracy of gene prioritization.

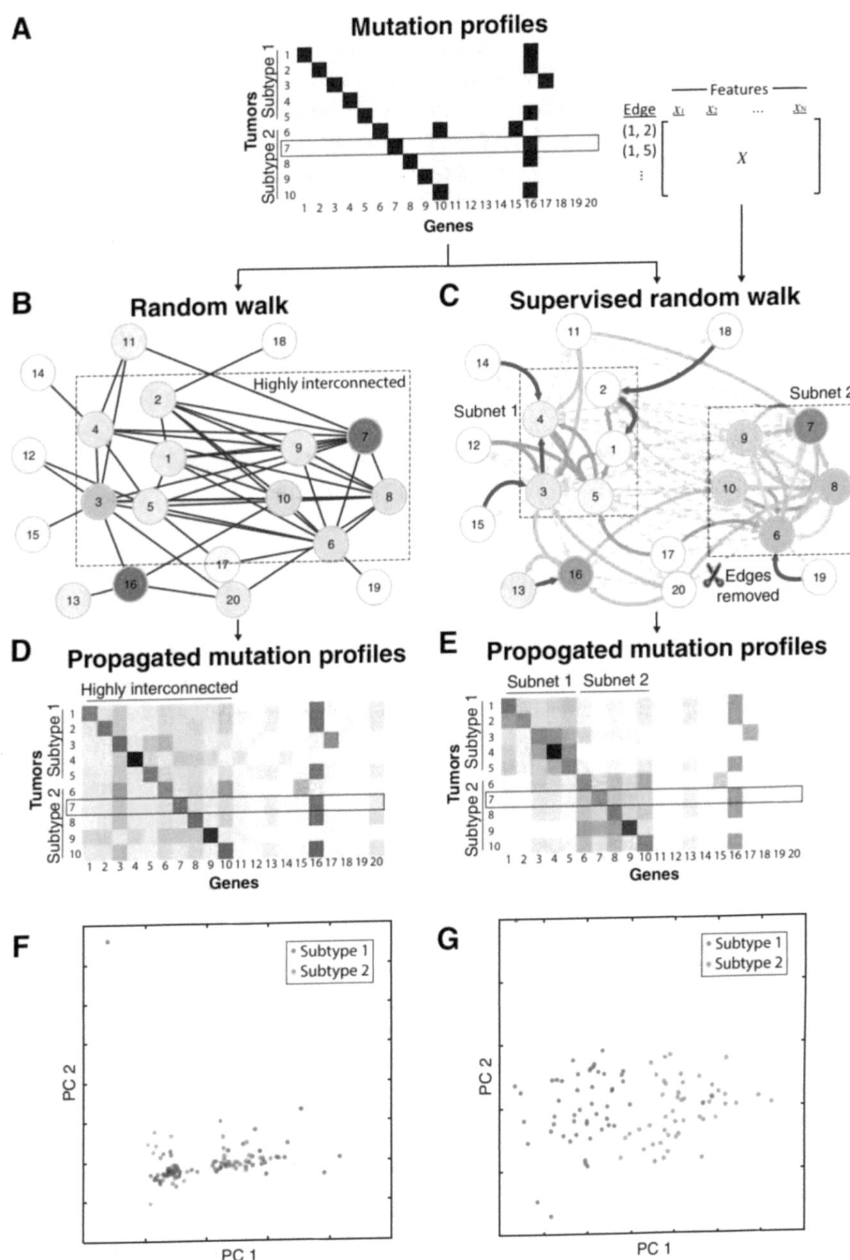

Fig. 9.2 Altered network after supervised network propagation. Unlike normal propagation in *B*, network edges from the supervised network in *C* are altered so that the subnetwork can represent the tumor-specific characteristics. Using propagated mutation profile from supervised network propagation, tumor subtypes are distinctly discernible as *G* shows (Reproduced from Zhang et al. 2018) [5]

Network propagation is a powerful tool that integrates gene expression data and gene network information for gene prioritization analysis [22, 23]. Network propagation propagates importance of a set of genes, called seeds, to the neighbor genes along the gene network. As a result of the network propagation, we obtain refined importance, which is basically the amount of effects that seeds have on all other genes. This refined importance is used for gene prioritization, namely, genes that have high refined importance are genes that are highly affected by seeds and vice versa. The basic application of network propagation for gene prioritization uses differentially expressed genes or differential expression value of genes for importance of seeds. To propagate differential expression-based importance, many currently available gene prioritization tools such as PRINCE [24], CATAPULT [25], DADA [26], Exome Walker [27], GUILD [28], PRINCIPLE [29], and ToppGene [30], use random walk with restart (RWR) [31] for the underlying algorithm.

A recent interesting research conducted by Ahn et al. [32] used network propagation method for transcription factor (TF) prioritization and TF target subnetwork construction on the time domain. Transcription factors (TFs) regulate expressions of down-stream genes in a cell. The regulatory relationships between TFs and target genes are represented in a graph model, which is called a *TF gene regulatory network (or TF network)*. Since a TF is a key cellular regulator to control multiple down-stream genes, researchers have made great efforts to elucidate the TF network. Ahn et al.'s study investigated plant response to environmental stress over time by characterizing the dynamics of TFs. This was done by exploiting the power of network propagation on TF networks.

9.4.2 PropaNet: TF Prioritization and Subnetwork Construction on Time-Series Gene Expression Data

PropaNet studies of the dynamics of transcription factors under a condition of interest by solving the problem of TF prioritization using network analysis. PropaNet takes three types of input data, time-series gene expression data that are measured at multiple time points, D, a template TF network, G, and a set of target genes of interest, S, and outputs a set of major regulatory transcription factors (TFs) for the target genes and their time-varying network at each time point. PropaNet operates in three steps as below and the process is visualized in Fig. 9.3.

- **Step 1. Instantiation of time-specific TF networks.** PropaNet instantiates time-specific networks from the given template network, G. PropaNet does this by including all TFs and differentially expressed target genes while filtering out unrelated genes. For each time point, the list of given target genes and differentially expressed genes are compared to include only the intersecting genes and discard others from the TF template network, $S \cap DEG$. S denotes the set of target genes, and DEG is the set of differentially expressed genes at

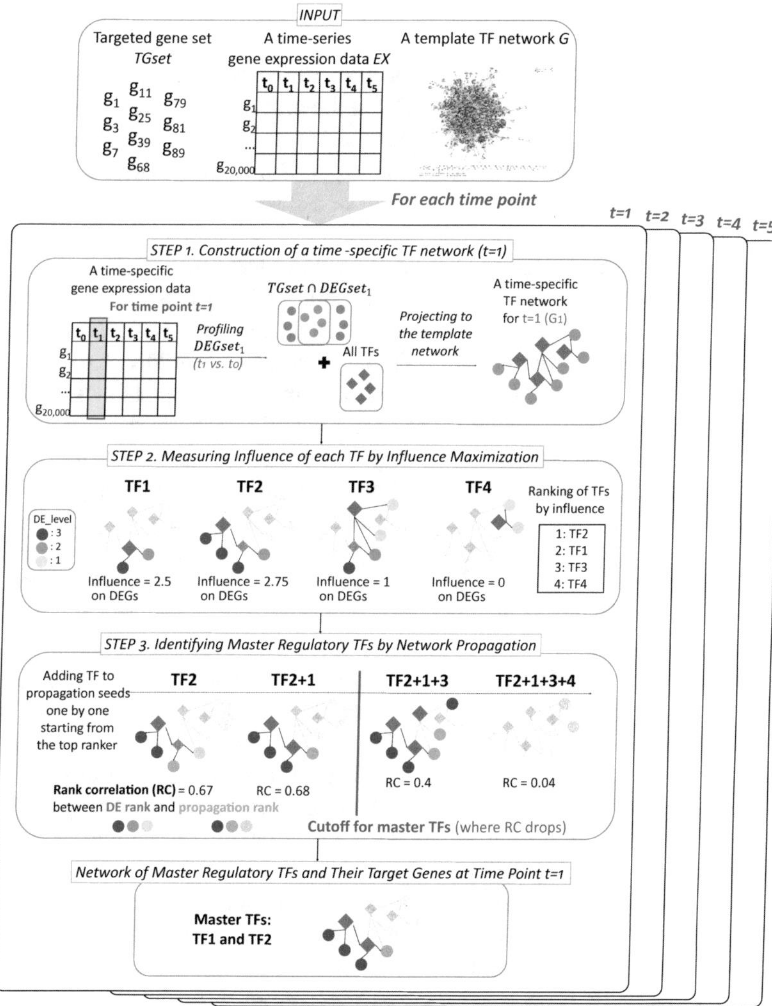

Fig. 9.3 Workflow of PropaNet. The PropaNet analysis takes three types of input data: time-series gene expression data that are measured at multiple time points, a template TF network, and a set of target genes. The goal of the PropaNet is to elucidate time-varying networks at each time point. It uses the influence maximization technique to produce a ranked list of TFs at each time point in the order of TF that explains DEGs better. Then, the network propagation technique is used to select a group of TFs that explains DEGs best as a whole. The process is done by iterating a network propagation simulation by adding a TF at a time, going down the list of TFs determined by the influence maximization technique [32]

time point t. DEG is determined by comparing gene expressions at time point $t \in \{1, \ldots, T\}$ with time $t = 0$.

- **Step 2. Time-specific measurement of the influence of each TFs by Influence Maximization.** PropaNet ranks the TFs in the order of influence to $S \cap DEG_t$ at each time point t through the Influence Maximization (IM) algorithm. IM is an algorithmic technique used in the network influence analysis to select a set of seed nodes to maximize the spread of influence (the expected number of influenced nodes) from a given network [33]. PropaNet uses a modified version of the labeled influence maximization algorithm [34]. The algorithm iteratively selects a TF, removes each edge with probability of $1 - p$, where p is the edge weight, counts reachable nodes, and computes the ratio of $S \cap DEG_t$ among the reachable nodes. A higher ratio of a TF indicates a higher influence on the DEG.
- **Step 3. Time-specific identification of major regulatory TFs by network propagation.** PropaNet simulates the TF-centered regulation process using the network propagation analysis. Each step of network propagation outputs ranked list of nodes in the network. The objective function, or the stopping criteria, is to find a ranked list of genes that are most similar to the ranked list of DEGs in terms of their p-values. This can be easily determined by computing Spearman's rank correlation coefficient (SCC) between the two ranked lists. More formally, the simulation is evaluated by the comparison of the ranking between the differential expression of observation $DE(v)$ and the inferred expression of network propagation $IP(v)$. This simulation is independently processed for each time point j. It first initializes the information of nodes, $IP(v)$ as 0 for $v \in V$. At time point j, we would have a list of TFs and their influence scores, $IL(t)$, that are measured in the previous step. It, then, initializes the most influencing TF (i.e., $\text{argmax}_t IL(t)$) as a set of seed S and conducts network propagation on the TF network G to update $IP(v)$. As the last step of the iteration, it measures the similarity of ranking, SCC, between $IP(v)$ and $DE(v)$. PropaNet then moves onto the next iteration by adding the next most influencing TF into the set of seeds S and then repeats the process of performing network propagation and computing SCC. After each iteration, PropaNet decides whether to accept the newly added TF or not. The TF is accepted if the SCC increases or declined otherwise. This process is continued through the list of the TFs until the coverage of the target genes exceeds half the number of DEGs at the time point. Finally, it produces S as a set of major regulatory TFs.

9.4.2.1 How Network Propagation Is Utilized Within PropaNet

The most basic network analysis considers only direct targets. However, network propagation along with Influence Maximization enables PropaNet to consider indirect regulatory power of a TF though multiple steps of transcription regulation. In addition, PropaNet takes into account of the regulatory power of multiple TFs simultaneously. Leveraging network propagation analyses, PropaNet detected known stress-responsive genes more accurately in terms of F1 scores than existing time series analysis methods [32].

On the other hand, there are limitations of PropaNet that have to be considered before analysis. PropaNet takes a template TF network as input and it is critical to use correct and comprehensive template network. In the case of PropaNet [32], PlantRegMap [35] was used as a template network, which was constructed by TF ChIP-seq experiments and the literature search. Thus, there is a possibility to include false TF-target interactions with respect to the specific condition of interest. However, under the assumption that the proportion of false TF-target interactions is small, the network propagation analyses can enhance signal-to-noise ratio for the true/false TF-target interactions by transmitting the information to indirect targets. The analysis results reported in [32] show that PropaNet determines the known stress response genes accurately; thus, the use of a template network seems to be empirically correct.

9.4.3 Differentially Expressed Gene (DEG) Detection

Differentially expressed gene (DEG) detection is a widely used analysis for relating a gene expression profile to phenotype. For example, let us assume that a different phenotype is observed when a cell is treated with a drug. Genes whose expressions are upregulated or downregulated compared to the control group are useful information to explain the phenotype difference. Since DEG detection is the most basic step in transcriptomic analysis, a number of DEG detection methods have been developed—such as limma [36], EBSeq [37], DESeq2 [38], edgeR [39], etc.

Although these DEG detection tools have effectively been used for numerous RNA expression data analysis tasks, those tools have some limitations. First, users have to set a cutoff value as a hyperparameter, which makes DEG detection arbitrary. If a strict cutoff value is given, relatively few DEGs would be detected, resulting in many false negatives. With a loose cutoff value, too many false positive DEGs will be determined. However, there is no good guideline for how to set the cutoff value. Moreover, DEG results from those tools vary a lot even for the same dataset because of the confounding factors (e.g., experimental conditions, interaction between genes, short read mapping tools, and data normalization methods) [40]. To address this issue, Moon et al. [40] proposed a machine learning-based DEG detection method, MLDEG.

The main problem is that many DEGs are not consistently detected by all DEG detection tools used. A gene may be detected as a DEG by some tools but may not be detected as a DEG by other tools. Therefore, combining the results of multiple DEG analyses would help decrease the variability of the results as well as increase accuracy. To combine DEG detection results by multiple DEG tools through a machine learning approach, it is necessary to prepare a training set of positive and negative examples. Positive examples should be true DEGs. Of course, we do not have this information. However, DEGs determined by many or all DEG detection tools are highly likely to be true DEGs, especially when determined with a stringent cutoff value. Negative examples or non DEGs can be determined easily since many genes are not determined by any of the DEG tools. With the condition-specific training data, we can easily build a classification model and use the model to determine which genes are DEGs. MLDEG uses features from the biological network and the expression profile of true DEGs in the training data. The features from true DEG are exploited to train a logistic regression model. Then, after training is done, the model classifies whether DEGs in the gray area are true DEG or not. MLDEG uses three types of features: expression feature, network property feature, and network propagation feature. The workflow of MLDEG is described in Fig. 9.4.

- First of all, *expression features* are the combined scores of the significance scores reported by each tools, e.g., p-value, and the amount of fold changes, log2FC.
- Second, *network property features* are the features extracted from the network properties of the condition-specific network. The condition-specific network can be instantiated by excluding the edges below a preset correlation score threshold from the template network. The template network used in the study is the PPI network from STRING database. The correlation score of each edge is calculated from the expression profile of two genes connected with an edge. The edge

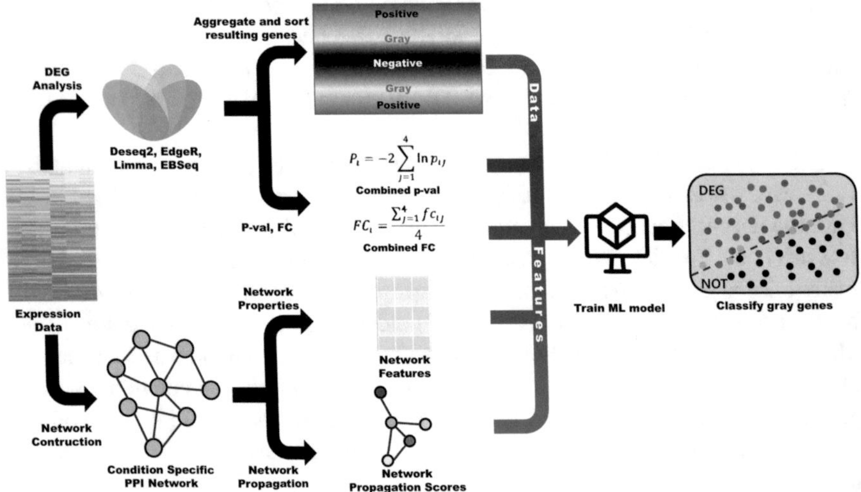

Fig. 9.4 The workflow of MLDEG. Training data are genes consistently detected as DEGs (positive data) or non-DEGs (negative data) from DEG detection tools used. From expression profile of data and a PPI network, features are extracted to train multiple logistic regression model. Network propagation feature is a feature that reflects the whole topology of a network. After a DEG classification model is trained, genes in the gray area are classified as DEGs or non-DEGs [40]

connecting two genes whose expressions are not highly correlated is trimmed to construct the condition-specific network.

- Last but not least, *network propagation feature* is the feature that reflects the whole topology of the network.

Here, we focus on the third feature to show how the network propagation is used to extract feature.

9.4.3.1 How Network Propagation Is Used to Extract a Feature for Machine Learning

Inspired by signed network propagation [41], Moon et al. initialized the network and propagate the seed values to extract network propagation feature. In the signed network propagation, it is assumed that a marker gene is exclusively expressed in control group or treatment group. Therefore, the class label is set to 1 for a treatment sample and -1 for a control sample when correlated with expression level of the marker gene. In other words, a gene with high correlation coefficient is the marker gene that discriminates the treatment sample from the control sample. The Pearson's correlation coefficient represents how much impact a gene has on the phenotype difference. If a gene is a marker gene, its absolute Pearson's correlation coefficient is close to one. If not, Pearson's correlation coefficient approaches to zero. These correlation coefficients of all genes are propagated as seeds to rank genes.

MLDEG takes the concept of how seeds are defined in the signed NP. When these correlation coefficients of each gene are propagated as seeds throughout the network using random walk with restart (RWR) algorithm, the propagated value of each gene represents the influence on other genes from seeds in the network. The propagated value is a *network propagation feature* of a gene that is used for logistic regression for classifying genes in the gray area. This feature reflects the whole network topology. Unlike conventional DEG detection tools where only the expression level difference of individual gene is exploited, MLDEG additionally considers relations among genes using the network propagation technique.

With other features—*expression feature* and *network property feature*—extracted from true DEGs and non-DEGs, a multiple logistic model is trained to classify DEGs in the gray area. When tested using 10 high-throughput RNA seq datasets, the performance of MLDEG was compared with the performances of four existing tools such as limma [36], EBSeq [37], DESeq2 [38], and edgeR [39]. MLDEG outperformed the other tools in terms of the number of DEGs detected compared to the ground truth genes that were reported in the original studies.

9.4.4 Hypothesis Testing

Transcriptome data, especially gene expression alteration at transcript level, can be very useful to investigate the association of phenotypes in specific experimental conditions. When there are multiple competing hypotheses on the phenotypic difference, network propagation can be useful to rank and select the most probable hypothesis using the network propagation technique with gene expression profile. For example, the causes of chronic graft loss after pancreatic islet transplantation are well defined. Among the plausible hypotheses are a higher rate of apoptosis due to ER stress [42, 43], hypoxia in end-portal venules in the liver [44, 45], and recurrent autoimmunity [46]. In addition, there are evidences that metabolic deterioration due to lipid accumulated around the islets (lipotoxicity) [47, 48] and toxicity of immunosuppressive drugs [49, 50] can result in graft loss. Also, insufficient immune suppression could also be an important cause of chronic islet loss. Among the multiple hypotheses, identifying the most probable hypothesis accompanying with the perturbations in gene expression is crucial.

9.4.4.1 How Network Propagation Is Utilized to Evaluate Hypotheses

To evaluate hypotheses using gene expression profile, one of the plausible methods is to design and perform a network propagation-based computational experiment. The rationale behind the experiment is as follows: if the hypothesis under consideration is the cause of a certain phenotype, then genes related to the hypothesis should "explain" the global gene expression well since the global gene expression profile can be a signature of the phenotype. Then, the main question is, how do we know if genes related to the hypothesis "explain" the global gene expression profile. This can be done using the network presentation as seeds. The evaluation process is as follows and the overview is shown in Fig. 9.5:

- Step 1. Genes relevant to each hypothesis are used as seed genes. The genes are collected from the literature search and the domain knowledge. Each hypothesis then can be represented by a set of seed genes.
- Step 2. Gene expression profile is used to calculate DEGs, and genes are ranked in terms of the level of differential expression in descending order.
- Step 3. A protein–protein interaction (PPI) network is built, and the seed genes are mapped to the network.
- Step 4. Network propagation is carried out on the network, and it yields the ranking of genes by the influences that the genes received from the seed genes.
- Step 5. The correlation between the two rankings is calculated using Pearson's correlation.

In summary, it is to compare the network propagation simulation with the hypotheses and the measured gene expression profiles in terms of gene rankings. The hypothesis with the highest coefficient is the most probable one among all

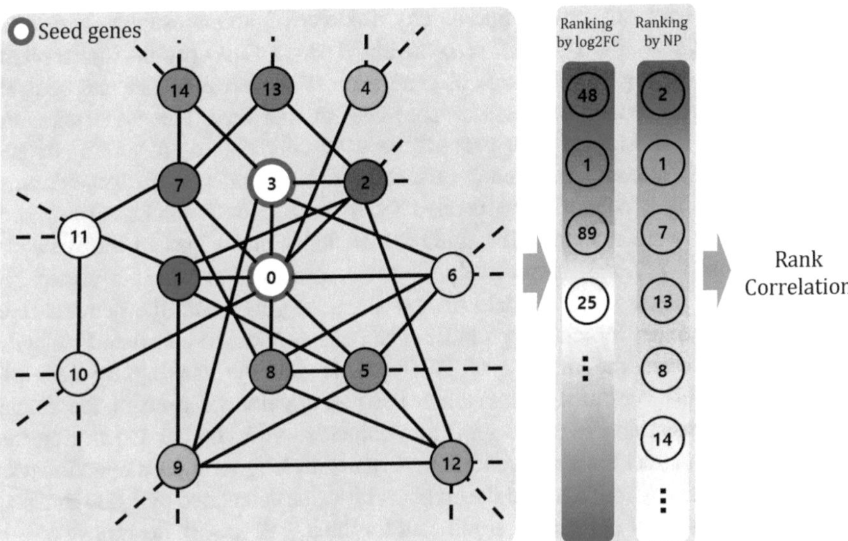

Fig. 9.5 The global effect of seed genes is measured by network propagation, and the rankings of the genes in a network are calculated in terms of how the genes are affected. Pearson's correlation is calculated between the ranking by network propagation and the ranking of DEG profile [51]

hypotheses, explaining the cause of the target phenotype. There is one more issue that needs to be considered. Since we are comparing ranking of a large number of genes, the correlation coefficient is small even for the best matching hypothesis. To complement this problem, random simulations are carried out 1,000 times to calculate the empirical p-value for each hypothesis to test the significance of the coefficients as shown in the equation below:

$$p^i = \frac{1}{N}\sum_{j=1}^{N}\begin{cases}1 \text{ if } c_{ij} > c_i^R\\ 0 \text{ otherwise.}\end{cases} \tag{9.7}$$

Here, i indicates each hypothesis and it ranges from 1 to the number of possible hypotheses. p^i indicates the empirical p-value of i-th hypothesis. N is the number of random simulation and it is 1,000. j indicates the j-th random simulation. c_{ij} is the coefficient of j-th random simulation of i-th hypothesis. c_i^R is the reference coefficient of i-th hypothesis.

The concept of hypothesis evaluation using network propagation was applied to a pig-to-nonhuman primates (NHP) xenotransplantation study to evaluate the hypotheses related to chronic islet graft rejection and the cause. As mentioned in the previous section, there are five hypotheses: ER stress [42, 52–54], islet exhaustion [55], lipotoxicity [47, 56–58], toxicity of immunosuppressant [49], and long-term graft rejection. To evaluate the hypotheses, seed genes were collected from the literature search. The numbers of seed genes were ten, nine, eight, nine, and ten for ER stress, islet exhaustion, lipotoxicity, toxicity of immunosuppressant, and long-term graft rejection (LTGR), respectively. Then, a DEG profile was created by calculating the log2 fold change of gene expression between case and control samples. Genes of which expression value was smaller than 1 in either case or control sample were removed to prevent the extremely high or low fold change calculated from the comparison between small numbers. Then, a PPI network was constructed using a network downloaded from the STRING database [59], and the seed genes were mapped. The numbers of nodes and edges in the network were 6,780 and 117,963, respectively. Network propagation was performed on the network to measure the global effect of the seed genes, and the genes in the network were ranked for each hypothesis. The next step was to calculate Pearson's correlation between the ranking of DEG profile and the ranking by network propagation. This was to see how much the participation of gene in the actual biological process that drives the graft loss coincided with the perturbation in the gene expression under the given condition. As a result, long-term graft rejection was the most probable hypothesis, and the empirical p-value calculated by 1,000 times of random simulations was the most significant for the hypothesis. In the biopsy result, it was observed that $CD3^+$ T cell heavily infiltrated the site of transplantation and long-term graft rejection was strongly correlated to the result.

9.4.5 Comparing Multiple Experiments

As mentioned in Sect. 9.4.3, one of the common approaches to understanding the phenotypic differences between samples is to prioritize genes by comparing the lists of differentially expressed genes (DEGs). However, there are multiple lists of DEGs from multiple experiments, and the number of DEGs is large, hundreds to thousands. Thus, comparing DEGs as lists is not easy. To ease the difficulties above, Venn diagram is commonly used. Venn diagram is a simple, yet powerful tool that can illustrate the portion of each gene set. It helps researchers to understand the common and distinctive characteristics of the experiments and assists in further investigation. However, there is a serious issue when Venn diagram is used to compare and analyze multiple experiments, each of which is performed in the control vs. treated setting. In general, each biological experiment is performed in different settings of control and the effect of treatment is unknown since it is the goal of the investigation from the experiment. A typical practice is to compare how many DEGs are common or unique in multiple experiments without considering the difference in control and treatment settings. Since Venn diagram is a very powerful tool, there are computational methods that are developed to utilize the power of Venn diagram [60–67]. However, none of these tools or systems consider the difference in control and treatment settings. To compensate the difference in control and treatment settings, some adjustments should be made using computational methods. Hur et al. [68] used a gene prioritization method to address this issue. Gene prioritization is a widely used strategy that can effectively rank genes. Among various strategies of prioritizing genes, network propagation is one of the widely used techniques that computes the influence of initial nodes (or seeds) to other nodes [1], and that can prioritize genes in the context of biological networks [18, 24, 27, 31, 69–71]. Gene prioritization that uses network propagation can be especially useful when there are multiple DEG lists that have different measurements. Network propagation can rank genes with a criterion of *network topology* instead of considering multiple expression values from DEG lists. However, network propagation methods require *seeds* as input, and the correct selection of seed genes is very important when prior knowledge is not available or is not enough to select seeds. Thus, domain experts need a way to explore network propagation freely with different seed selections. To address this technical issue, Venn-diaNet: a web-based Venn diagram-based network analysis framework that can prioritize genes to compare multiple biological experiments of transcriptome data was developed [68]. Venn-diaNet assumes that each area of the Venn diagram represents a subset of DEGs with a specific biological meaning. These subsets of DEGs can be used as a guidance to logically select seeds for the subset of genes that the user is interested in.

9.4.5.1 How Network Propagation Is Utilized in Venn-diaNet

Venn-diaNet is designed to offer a convenient web-based user interface where Venn diagrams of DEGs is automatically computed and network propagation analysis can be performed to investigate which genes are relevant to certain phenotypes. Venn-diaNet operates in four major steps to prioritize genes and compare multiple biological experiments as described in Fig. 9.6.

- **Step 1. Take multiple DEG lists as input from the user.**
- **Step 2. Generate and illustrate Venn diagram from the DEG lists.** Venn-diaNet considers each experiment as a set for the diagram. Therefore, With the given number (=n) of experiments E, Venn-diaNet generates a diagram of n circles that have a maximum of $2^n - 1$ regions. Each region is denoted as C_i $(1 \leq i \leq 2^n - 1)$, while each C_i contains genes of

$$\mathbf{C}_i = \{\mathbf{g} : \mathbf{g} \in \textstyle\bigcap_{j=1}^{N} \mathbf{G}(\mathbf{b}_j)\} \tag{9.8}$$

$$\mathbf{G}(\mathbf{b_j}) = \begin{cases} E_j & if \;\; j = 1 \\ E_j^c & if \;\; j = 0, \end{cases} \tag{9.9}$$

where b represents a binary number of C_i (i.e., $C_1 = 001$), while b_j indicates the position of digits (i.e., $b_1 = 1$, $b_2 = 0$, and $b_3 = 0$). If Venn-diaNet receives DEG lists from 3 experiments, then Venn-diaNet illustrates a Venn diagram of 3 sets (E_1,E_2,E_3) that have 7 regions (C_1, C_2, C_3, $\cdots$ C_7), where C_7 contains genes of $E_1 \cap E_2 \cap E_3$.

- **Step 3. Propagate seeds that are selected by a user on a network.** A user can select multiple (or a single) C_i as seeds. As the network propagation analysis results will vary on its seed selection, it is important to select specific C_i that matches to the user's interest. Venn-diaNet will rank genes that are similar (or dissimilar) to the selected seed genes. Therefore, if the user is interested in ranking DEGs in C_i, the user should select C_j $(j \neq i)$ that is functionally or biology close (or far) from C_i.
- **Step 4. Perform network propagation and rank DEGs.** When a set of seed DEGs are selected, Venn-diaNet instantiates a protein–protein interaction (PPI) network of DEGs from STRING DB [72] and performs network propagation with a Markov random walk with restart algorithm [73] using the seeds selected in the previous step. To be specific, Venn-diaNet used binary values (1 or 0) to initialized seed genes p^0 and represents an unweighted graph as an adjacency matrix A. When the algorithm stops, Venn-diaNet returns a ranked gene list based on the network propagation analysis result. This way, Venn-diaNet quantifies the influence of seed DEGs to the remaining DEGs.

Venn-diaNet was tested whether Venn-diaNet can reproduce the results reported in the original studies [74, 75]. Since the propagation result differs a lot depending on the initial seed values, the seed selection is a critical factor for Venn-diaNet. Especially in Venn-diaNet, how to choose seeds determines the objective of the experiment. Thus, several guidelines on how to select proper seed genes were proposed as below.

The first suggestion is to consider *condition-specific genes* as seeds. For example, given two different lists of DEGs from tissue A and tissue B, respectively, let us assume that the objective here is to prioritize tissue A-specific DEGs that have a similar function to the tissue B-specific DEGs. In this case, tissue B-specific DEGs are used as seeds to retrieve A-specific DEGs that are highly related to B-specific DEGs.

The second scenario is to consider *intersecting genes* of experimental conditions as seeds. If the user is interested in tissue A-specific DEGs that have similar function among two tissues, the intersection between two tissues can be used as seeds.

9.5 Transcriptome Data Analysis at Pathway Level

A biological pathway is a series of interactions among molecules in a cell that leads to a certain product or a change in a cell [76].

A pathway can be viewed as abstraction of high-dimensional molecular entities such as genes, proteins, and metabolites. Using well-curated pathway databases such as KEGG [77], conventional pathway analysis focuses on inferring pathway activations with the molecular measurement of a cell. However, the limitation of these approaches is that it does not consider relationship among pathways. In this section, we will introduce a novel method that constructs the pathway–pathway

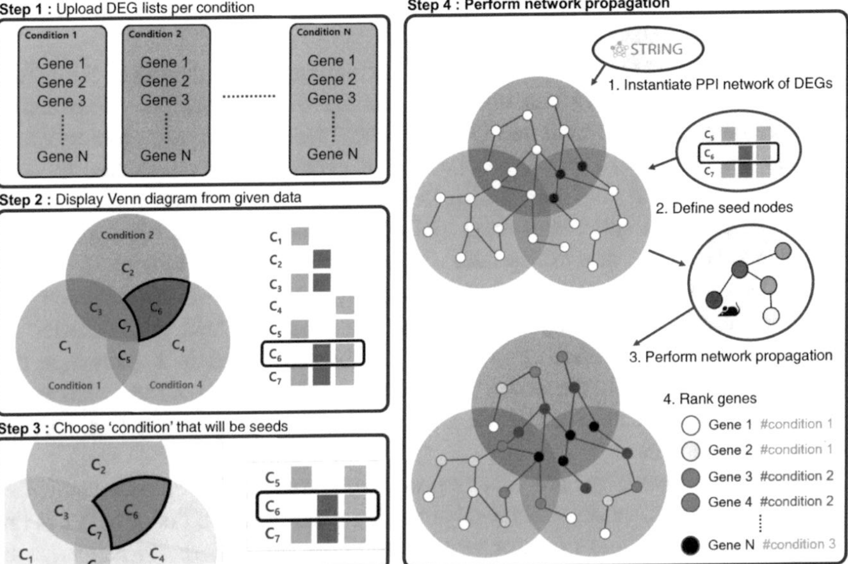

Fig. 9.6 Venn-diaNet workflow. *Step 1*: Venn-diaNet, as input, takes DEG lists per experiment from user. *Step 2*: Uploaded DEGs from Step 1 are interpreted with a Venn diagram and they are organized as sets in a table. Now, a user can repeat Steps 3 and 4 as many times as she or he wants. *Step 3*: One or multiple regions of Venn diagram, C_i, are selected as seeds for further network propagation analysis. *Step 4*: Once the seed is selected, Venn-diaNet instantiates a PPI network of DEGs from STRING DB. Network propagation with given seeds from the previous steps. As a result, DEGs are ranked by the probability score calculated during the Markov Random Walk

network [78] and discuss on how the network propagation is leveraged in that network [79].

9.5.1 *Disease Subtype Classification with Pathway Information*

According to the report of GLOBOCAN 2018, cancer is a main cause of death worldwide and there will be estimated 9.6 million cancer deaths in 2018 [80]. One of the reasons cancer is a fatal disease is cancer heterogeneity. Even though patients suffer from a same cancer, due to the cancer heterogeneity, they show different prognosis on the same treatments. Considering this property, in medical fields, different therapies are performed according to the subtypes. For example, node negative and estrogen receptor-positive breast cancers such as luminal type are treated by tamoxifen [81]. In case of HER2 positive metastatic breast cancer, Herceptin (Trastuzumab) is widely used for preventing malfunction of over-expressed HER2 [82]. Therefore, many researchers investigate subtypes of cancer on various data and methods [83–87] as well as mutation-based classification [5] described in Sect. 9.3.1.

Gene expression data is commonly used for defining cancer subtypes. To analyze gene expression data, we can utilize an additional biological prior knowledge, called as pathway, to support and improve the subtype classification performances and interpretations. Pathway is a collection of genes that are considered as same or similar biological functions and it is represented in graph form. Actually, in living organisms, a single and huge biological process controls life cycle of the organisms. However, due to huge scale and complexity of the biological process, it is not easy to interpret as a whole. For many decades, to investigate complex biological mechanisms, many biologists and medical researchers have dissected the biological process into smaller, more biologically meaningful units, pathways, that are easy to understand. Therefore, pathways are valuable domain knowledge that contains interactions between genes, and they can be clues of elucidating complex mechanisms of cancer.

Even though pathways are valuable prior knowledge, there are two challenges for classification of cancer subtypes by utilizing pathways. The first problem is that pathways are commonly used for visualizing biological process, so pathways are not designed as computable resources. The second problem is that pathways are artificial dissections of the whole biological process, each of which is designed to show specific biological processes such as cell cycle and apoptosis. However, to investigate the whole biological process, it is necessary to integrate the dissected pathways.

To address these issues, Lee et al. [79] proposed an attention-based ensemble model of graph convolutional network (GCN) model for cancer subtype classification tasks on the gene expression data and pathways. To model the pathways in the graph forms, GCN was utilized to extract gene interactions in the pathway with gene expression data. More specifically, they built a GCN model on each pathway in the KEGG database [77]. And then, to overcome partiality of pathways, described as the second challenge, prediction results of those GCN models were integrated using attention-based ensemble model. To improve interpretability of the model, authors used attention weight to explain which pathways are important on the given patient data. Along with attention model, network propagation algorithm was used for bridging the gap between pathways and biological regulation mechanisms by focusing on the transcription factors (TFs) that regulates expression levels of target genes (Fig. 9.7).

9.5.1.1 What Is the Attention-Based Ensemble Model of Graph Convolutional Networks

To help the reader to understand, we first discuss how to extract gene interactions in a pathway and how to aggregate information of multiple pathways. Given a single pathway p, the pathway is represented as a graph $G^p = (V^p, E^p)$, where V^p is the gene set and E^p is a collection of gene interactions. Based on the graph G^p and

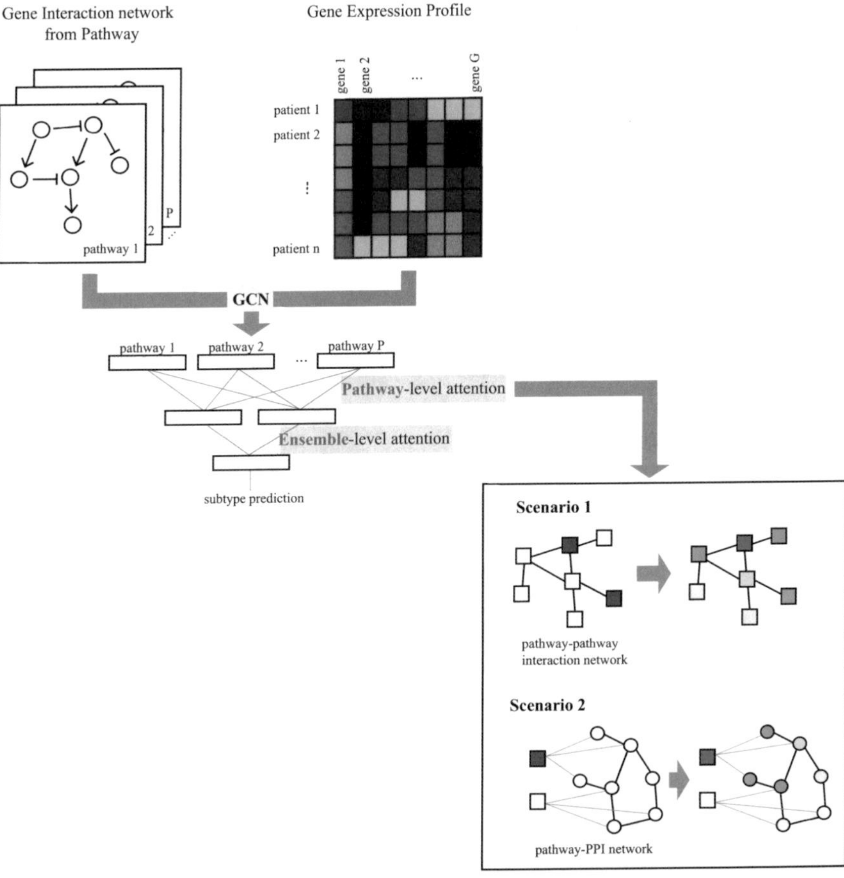

Fig. 9.7 The GCN and attention-based ensemble model and pathway-level network propagation. Given expression data and pathways, gene interactions in the pathway are captured by graph convolutional network. To reconstruct a whole biological process, extracted information from each pathway is aggregated by attention-based ensemble model. Based on the pathway attention scores from pathway-level attention, pathway-level network propagation is performed on the patient-specific pathway interaction network

gene expression matrix X, Lee et al. utilized a graph convolutional network (GCN) to capture localized gene expression patterns considering the graph topology, i.e., relationships of gene interactions.

In case of conventional convolutional network, input images are represented as a grid structure and neighbors of a specific pixel are easily obtained. Similar with the convolutional network, GCN also focuses on the neighbors of a certain node. However, unlike the grid structure, neighbor information on the graph are the needed additional procedure. Briefly, GCN utilizes a graph topology from the Laplacian matrix $L^p = D^p - A^p$ of the graph. Here, D^p is a weighted degree matrix of G^p

and A^p is an adjacency matrix of G^p. By performing an eigenvalue decomposition on the Laplacian matrix $L^p = U \Lambda U^T|_p$, graph convolution will be performed as

$$L^p_{spectral} = U g_\theta(\Lambda) U^T X|_p, \tag{9.10}$$

where $g_\theta(\Lambda)|_p$ is a polynomial convolution filter that is applied on the diagonal matrix Λ^p.

$$g_\theta(\Lambda)|_p = \sum_{k=0}^{K-1} \theta_k \Lambda^k|_p \tag{9.11}$$

In practice, many approximation techniques are applied on the convolutional filters because $g_\theta(\Lambda)|_p$ takes $O(n^2)$ to calculate the polynomial filter. If readers want to know more details, please see [88, 89]. Here, GCN is utilized to transform a pathway graph, which is a gene–gene interaction network in pathway, into a d-dimensional vector.

After the GCN pathway models are trained for P pathways, reconstruction of a whole biological process is done by aggregating extracted pathway information. To improve interpretability of deep learning models, authors combined multiple GCN models based on the attention-based ensemble scheme. Attention is basically deep learning-based weighting scheme among features, which are pathways in this research.

Here is precise representation of what was conducted on [79]. From GCN step, gene expression data X_i for patient i are transformed into P number of pathway encoding vectors $h^p(X_i)$, where $p = 1, 2, \ldots, P$, and those encoding vectors are concatenated into a large matrix form $h(X_i) \in \mathbb{R}^{P \times d}$, where d is size of encoding. On the matrix $h(X_i)$, attention scores for each pathway are calculated through the neural network, where there are a nodes in hidden layer when computing attention score. Details on how attention scores are retrieved are as follows:

$$\begin{aligned}
&W \in \mathbb{R}^{d \times a}, b \in \mathbb{R}^a, u \in \mathbb{R}^a \\
&Y = tanh(h(X_i)W + b) \in \mathbb{R}^{P \times a} \\
&\alpha = Softmax(Yu) \in \mathbb{R}^P \\
\tilde{h}(X_i)_j = &\sum_{j=1}^{P} h(X_i)_j \alpha_j \\
&\text{where } \tilde{h}(X_i) \in \mathbb{R}^d
\end{aligned} \tag{9.12}$$

The attention scores α mean that which pathways are important in classifying a subtype of given cancer patient. Based on this framework, multiple attention-based aggregations were designed to consider variances in a single patient, as known as intra-tumor heterogeneity. As a conclusion, two-level attention layers that one is pathway-level and the other is ensemble-level attention were utilized for better cancer subtype classifications.

9.5.1.2 How Network Propagation Is Utilized in the Attention-Based Ensemble Model of Graph Convolutional Networks

From the GCN and attention-based ensemble model, attention weights for pathways are easily obtained. To interpret biological meaning of attention weights for pathways, the network propagation technique was used. Unlike network propagation approaches described in previous sections, seed nodes on the network are pathways, not a set of genes. Considering this difference, there are two possible research scenarios for the pathway-level network propagation. The first one is that mining significant pathway subnetworks on the pathway interaction network. The second one is that finding regulators related with highly focused pathways from attention scores.

In case of first approach, key point is how to construct pathway interaction networks. Moon et al. [78] proposed condition-specific pathway interaction networks using gene expression data and protein–protein interaction (PPI) network. Activation scores of pathway interaction are calculated with PPI subnetworks induced by overlapping genes of two pathways. Using [78], a patient-specific pathway interaction network is created. On the patient-specific network, seed nodes or values are assigned by top-ranked pathways from attention scores or as a whole attention scores. After selecting valuable seeds, network propagation is performed and pathway subnetworks are also recruited as similar way of previous studies [4, 5]. As a result of diffusing pathway attention scores across the pathway interaction network, we can identify pathways having small attention weights but highly interact with others. For example, the GCN & attention-based ensemble model was utilized to analysis subtype classification of TCGA BRCA data (The Cancer Genome Atlas, Breast invasive carcinoma).

In this analysis, interesting pathway rescued by the network propagation was "Retrograde endocannabinoid signaling (hsa04723)." The endocannabinoid system was related with tumor growth inhibition and/or cancer cell death, and it interacted with a variety of biological pathways such as apoptosis and autophagy [90]. This pathway ranked low in terms of attention weights on all patients. However, the pathway was identified as a significant pathway by the network propagation analysis that considered cross-talks with other important pathways such as Wnt signaling pathway and HIF-1 signaling pathways.

In case of second approach, key point is how to link pathways and genes and how to represent these relations on a single integrated network. The simplest way is to construct a network that connects the genes belonging to a pathway with the corresponding pathway node. Seeds in the configured network can be selected in the same way as the first approach. If network propagation is used in this way, the network itself is not condition-specific because it uses public information. But the advantage is that we can find regulators such as TFs that are tightly cooperated with pathways [91]. These analyses can identify TFs that mostly affect regardless of the subtype of the disease, and further analyzes can identify subtype-specific TFs. By analyzing the pathways associated with the excavated TFs, we can find abnormal biological functions that are considered as clues to treat the disease.

9.6 Conclusion

Network analysis is an effective and useful method for investigating interactions among multiple entities. Among many network analysis methods, network propagation has recently been used extensively. While most network analysis methods are static, network propagation is dynamic in that it can capture time information and can examine intermediary steps of the flow of influence to better understand the overall interactions within the network. In addition, unlike the existing methods that only considers parts of the network, network propagation can take the whole network topology into consideration when studying the relationships among the nodes. Network propagation is particularly useful for the analysis of molecular biology data that have high complexity of interactions between numerous biological entities such as genes or proteins within various biological phenomena such as outbreak of a disease or stress response. We introduced in this chapter how recent studies utilize network propagation techniques in analyzing multi-omics data at different levels of complexity. At DNA level, pan-cancer mutation data can be propagated based on the somatic mutation profiles of SNVs and CNAs in order to detect subnetworks that contain rare somatic mutations and classify tumor subtypes using genome information only. At RNA level, transcriptome data can be propagated to prioritize genes with respect to certain phenotypes. At pathway level, network propagation can be used along with attention model to improve the interpretability of a GCN model for cancer subtype classification based on gene expression data and pathway information. Not only is network propagation a powerful network analysis tool on its own but it can even be exploited further by integrating with other tools due to its high flexibility and applicability. As we surveyed in this chapter, network propagation can be used for a wide range of applications such as testing of multiple hypotheses, extracting feature to be utilized in another model or provision of better interpretation of the result. Even though network propagation is very useful and powerful method for investigating complex relationships among entities, it does have limitations. One of the limitations would be the weak statistical meaning of the raw propagation scores, which is often not enough information to draw conclusions of research investigations. In addition, seeds of the network propagation need to be defined by prior knowledge most of the time. This may lead to biased outcomes that favor well-studied and known nodes [2]. Many computational scientists are working on overcoming the shortcomings of network propagation [26, 92], and advances on network propagation algorithms will make network propagation useful for answering challenging research questions across different fields.

References

1. Cowen, L., Ideker, T., Raphael, B.J., Sharan, R.: Network propagation: a universal amplifier of genetic associations. Nat. Rev. Genet. **18**(9), 551 (2017)
2. Biran, H., Kupiec, M., Sharan, R.: Comparative analysis of normalization methods for network propagation. Front. Genet. **10**, 4 (2019)
3. Stratton, M.R., Campbell, P.J., Futreal, P.A.: The cancer genome. Nature **458**(7239), 719 (2009)
4. Leiserson, M.D., Vandin, F., Wu, H.T., Dobson, J.R., Eldridge, J.V., Thomas, J.L., Papoutsaki, A., Kim, Y., Niu, B., McLellan, M., et al.: Pan-cancer network analysis identifies combinations of rare somatic mutations across pathways and protein complexes. Nature Genetics **47**(2), 106 (2015)
5. Zhang, W., Ma, J., Ideker, T.: Classifying tumors by supervised network propagation. Bioinformatics **34**(13), i484–i493 (2018)
6. Vogelstein, B., Papadopoulos, N., Velculescu, V.E., Zhou, S., Diaz, L.A., Kinzler, K.W.: Cancer genome landscapes. Science **339**(6127), 1546–1558 (2013)
7. Vandin, F., Clay, P., Upfal, E., Raphael, B.J.: Discovery of mutated subnetworks associated with clinical data in cancer. In: Biocomputing 2012, pp. 55–66. World Scientific (2012)
8. Grasso, C.S., Wu, Y.M., Robinson, D.R., Cao, X., Dhanasekaran, S.M., Khan, A.P., Quist, M.J., Jing, X., Lonigro, R.J., Brenner, J.C., et al.: The mutational landscape of lethal castration-resistant prostate cancer. Nature **487**(7406), 239 (2012)
9. Ye, J., Pavlicek, A., Lunney, E.A., Rejto, P.A., Teng, C.H.: Statistical method on nonrandom clustering with application to somatic mutations in cancer. BMC Bioinformatics **11**(1), 11 (2010)
10. Ryslik, G.A., Cheng, Y., Cheung, K.H., Modis, Y., Zhao, H.: Utilizing protein structure to identify non-random somatic mutations. BMC Bioinformatics **14**(1), 190 (2013)
11. Martincorena, I., Campbell, P.J.: Somatic mutation in cancer and normal cells. Science **349**(6255), 1483–1489 (2015)
12. Greenman, C., Stephens, P., Smith, R., Dalgliesh, G.L., Hunter, C., Bignell, G., Davies, H., Teague, J., Butler, A., Stevens, C., et al.: Patterns of somatic mutation in human cancer genomes. Nature **446**(7132), 153 (2007)
13. Hofree, M., Shen, J.P., Carter, H., Gross, A., Ideker, T.: Network-based stratification of tumor mutations. Nature Methods **10**(11), 1108 (2013)
14. Wang, S., Ma, J., Zhang, W., Shen, J.P., Huang, J., Peng, J., Ideker, T.: Typing tumors using pathways selected by somatic evolution. Nature Communications **9**(1), 4159 (2018)
15. Brennan, C.W., Verhaak, R.G., McKenna, A., Campos, B., Noushmehr, H., Salama, S.R., Zheng, S., Chakravarty, D., Sanborn, J.Z., Berman, S.H., et al.: The somatic genomic landscape of glioblastoma. Cell **155**(2), 462–477 (2013)
16. Verhaak, R.G., Hoadley, K.A., Purdom, E., Wang, V., Qi, Y., Wilkerson, M.D., Miller, C.R., Ding, L., Golub, T., Mesirov, J.P., et al.: Integrated genomic analysis identifies clinically relevant subtypes of glioblastoma characterized by abnormalities in pdgfra, idh1, egfr, and nf1. Cancer Cell **17**(1), 98–110 (2010)
17. Prat, A., Perou, C.M.: Deconstructing the molecular portraits of breast cancer. Molecular Oncology **5**(1), 5–23 (2011)
18. Lee, I., Blom, U.M., Wang, P.I., Shim, J.E., Marcotte, E.M.: Prioritizing candidate disease genes by network-based boosting of genome-wide association data. Genome Research **21**(7), 1109–1121 (2011)
19. Li, Y., Li, J.: Disease gene identification by random walk on multigraphs merging heterogeneous genomic and phenotype data. In: BMC Genomics, vol. 13, p. S27. BioMed Central (2012)
20. Isik, Z., Baldow, C., Cannistraci, C.V., Schroeder, M.: Drug target prioritization by perturbed gene expression and network information. Scientific Reports **5**, 17,417 (2015)

21. Ma, X., Lee, H., Wang, L., Sun, F.: Cgi: a new approach for prioritizing genes by combining gene expression and protein–protein interaction data. Bioinformatics **23**(2), 215–221 (2006)
22. Nitsch, D., Tranchevent, L.C., Thienpont, B., Thorrez, L., Van Esch, H., Devriendt, K., Moreau, Y.: Network analysis of differential expression for the identification of disease-causing genes. PLoS ONE **4**(5), e5526 (2009)
23. Nitsch, D., Gonçalves, J.P., Ojeda, F., De Moor, B., Moreau, Y.: Candidate gene prioritization by network analysis of differential expression using machine learning approaches. BMC Bioinformatics **11**(1), 460 (2010)
24. Vanunu, O., Magger, O., Ruppin, E., Shlomi, T., Sharan, R.: Associating genes and protein complexes with disease via network propagation. PLoS Comput. Biol. **6**(1), e1000,641 (2010)
25. Singh-Blom, U.M., Natarajan, N., Tewari, A., Woods, J.O., Dhillon, I.S., Marcotte, E.M.: Prediction and validation of gene-disease associations using methods inspired by social network analyses. PLoS ONE **8**(5), e58,977 (2013)
26. Erten, S., Bebek, G., Ewing, R.M., Koyutürk, M.: Dada: degree-aware algorithms for network-based disease gene prioritization. BioData Mining **4**(1), 19 (2011)
27. Smedley, D., Köhler, S., Czeschik, J.C., Amberger, J., Bocchini, C., Hamosh, A., Veldboer, J., Zemojtel, T., Robinson, P.N.: Walking the interactome for candidate prioritization in exome sequencing studies of mendelian diseases. Bioinformatics **30**(22), 3215–3222 (2014)
28. Guney, E., Oliva, B.: Exploiting protein-protein interaction networks for genome-wide disease-gene prioritization. PLoS ONE **7**(9), e43,557 (2012)
29. Gottlieb, A., Magger, O., Berman, I., Ruppin, E., Sharan, R.: Principle: a tool for associating genes with diseases via network propagation. Bioinformatics **27**(23), 3325–3326 (2011)
30. Chen, J., Bardes, E.E., Aronow, B.J., Jegga, A.G.: Toppgene suite for gene list enrichment analysis and candidate gene prioritization. Nucleic Acids Res. **37**(suppl_2), W305–W311 (2009)
31. Köhler, S., Bauer, S., Horn, D., Robinson, P.N.: Walking the interactome for prioritization of candidate disease genes. Am. J. Hum. Genet. **82**(4), 949–958 (2008)
32. Ahn, H., Jo, K., Jung, D., Park, M., Hur, J., Jung, W., Kim, S.: Propanet: Time-varying condition-specific transcriptional network construction by network propagation. Front. Plant Sci. **10**, 698 (2019)
33. Li, Y., Fan, J., Wang, Y., Tan, K.L.: Influence maximization on social graphs: A survey. IEEE Trans. Knowl. Data Eng. (2018)
34. Li, F.H., Li, C.T., Shan, M.K.: Labeled influence maximization in social networks for target marketing. In: 2011 IEEE Third International Conference on Privacy, Security, Risk and Trust and 2011 IEEE Third International Conference on Social Computing, pp. 560–563. IEEE, Boston (2011). https://doi.org/10.1109/PASSAT/SocialCom.2011.152
35. Jin, J., Tian, F., Yang, D.C., Meng, Y.Q., Kong, L., Luo, J., Gao, G.: PlantTFDB 4.0: toward a central hub for transcription factors and regulatory interactions in plants. Nucleic Acids Res. **45**(D1), D1040–D1045 (2016). https://doi.org/10.1093/nar/gkw982
36. Ritchie, M.E., Phipson, B., Wu, D., Hu, Y., Law, C.W., Shi, W., Smyth, G.K.: limma powers differential expression analyses for rna-sequencing and microarray studies. Nucleic Acids Res. **43**(7), e47–e47 (2015)
37. Leng, N., Dawson, J.A., Thomson, J.A., Ruotti, V., Rissman, A.I., Smits, B.M., Haag, J.D., Gould, M.N., Stewart, R.M., Kendziorski, C.: Ebseq: an empirical bayes hierarchical model for inference in rna-seq experiments. Bioinformatics **29**(8), 1035–1043 (2013)
38. Love, M.I., Huber, W., Anders, S.: Moderated estimation of fold change and dispersion for rna-seq data with deseq2. Genome Biology **15**(12), 550 (2014)
39. Robinson, M.D., McCarthy, D.J., Smyth, G.K.: edger: a bioconductor package for differential expression analysis of digital gene expression data. Bioinformatics **26**(1), 139–140 (2010)
40. Moon, J.H., Lee, S., Pak, M., Hur, B., Kim, S.: MLDEG: A Machine Learning Approach to Identify Differentially Expressed Genes Using Network Property and Network Propagation. IEEE Transactions on Computational Biology and Bioinformatics (under review)

41. Zhang, W., Johnson, N., Wu, B., Kuang, R.: Signed network propagation for detecting differential gene expressions and dna copy number variations. In: Proceedings of the ACM Conference on Bioinformatics, Computational Biology and Biomedicine, pp. 337–344. ACM (2012)
42. Fonseca, S.G., Gromada, J., Urano, F.: Endoplasmic reticulum stress and pancreatic β-cell death. Trends Endocrinol. Metab. **22**(7), 266–274 (2011)
43. Negi, S., Park, S.H., Jetha, A., Aikin, R., Tremblay, M., Paraskevas, S.: Evidence of endoplasmic reticulum stress mediating cell death in transplanted human islets. Cell Transplantation **21**(5), 889–900 (2012)
44. Lau, J., Henriksnäs, J., Svensson, J., Carlsson, P.O.: Oxygenation of islets and its role in transplantation. Curr. Opin. Organ Transplant. **14**(6), 688–693 (2009)
45. Zheng, X., Wang, X., Ma, Z., Sunkari, V.G., Botusan, I., Takeda, T., Björklund, A., Inoue, M., Catrina, S., Brismar, K., et al.: Acute hypoxia induces apoptosis of pancreatic β-cell by activation of the unfolded protein response and upregulation of chop. Cell Death Dis. **3**(6), e322 (2012)
46. Pugliese, A., Reijonen, H.K., Nepom, J., Burke III, G.W.: Recurrence of autoimmunity in pancreas transplant patients: research update. Diabetes Management (London, England) **1**(2), 229 (2011)
47. Lee, Y., Ravazzola, M., Park, B.H., Bashmakov, Y.K., Orci, L., Unger, R.H.: Metabolic mechanisms of failure of intraportally transplanted pancreatic β-cells in rats: role of lipotoxicity and prevention by leptin. Diabetes **56**(9), 2295–2301 (2007)
48. Leitão, C.B., Bernetti, K., Tharavanij, T., Cure, P., Lauriola, V., Berggren, P.O., Ricordi, C., Alejandro, R.: Lipotoxicity and decreased islet graft survival. Diabetes Care **33**(3), 658–660 (2010)
49. Barlow, A.D., Nicholson, M.L., Herbert, T.P.: Evidence for rapamycin toxicity in pancreatic β-cells and a review of the underlying molecular mechanisms. Diabetes **62**(8), 2674–2682 (2013)
50. Drachenberg, C.B., Klassen, D.K., Weir, M.R., Wiland, A., Fink, J.C., Bartlett, S.T., Cangro, C.B., Blahut, S., Papadimitriou, J.C.: Islet cell damage associated with tacrolimus and cyclosporine: Morphological features in pancreas allograft biopsies and clinical correlation1. Transplantation **68**(3), 396–402 (1999)
51. Kim, H.J., Moon, J.H., Chung, H., Shin, J.S., Kim, B., Kim, J.M., Kim, J.S., Yoon, I.H., Min, B.H., Kang, S.J., et al.: Bioinformatic analysis of peripheral blood rna-sequencing sensitively detects the cause of late graft loss following overt hyperglycemia in pig-to-nonhuman primate islet xenotransplantation. Scientific Reports **9**(1), 1–11 (2019)
52. Rickels, M.R., Collins, H.W., Naji, A.: Amyloid and transplanted islets. New Engl. J. Med. **359**(25), 2729 (2008)
53. Potter, K., Abedini, A., Marek, P., Klimek, A., Butterworth, S., Driscoll, M., Baker, R., Nilsson, M., Warnock, G., Oberholzer, J., et al.: Islet amyloid deposition limits the viability of human islet grafts but not porcine islet grafts. Proc. Natl. Acad. Sci. **107**(9), 4305–4310 (2010)
54. Westermark, G.T., Westermark, P., Berne, C., Korsgren, O.: Widespread amyloid deposition in transplanted human pancreatic islets. New Engl. J. Med. **359**(9), 977–979 (2008)
55. Kim, J.W., Yoon, K.H.: Glucolipotoxicity in pancreatic β-cells. Diabetes Metab. J. **35**(5), 444–450 (2011)
56. Brown, M.S., Goldstein, J.L.: The srebp pathway: regulation of cholesterol metabolism by proteolysis of a membrane-bound transcription factor. Cell **89**(3), 331–340 (1997)
57. Brown, M.S., Goldstein, J.L.: Sterol regulatory element binding proteins (srebps): controllers of lipid synthesis and cellular uptake. Nutrition Reviews **56**(suppl_1), S1–S3 (1998)
58. Kakuma, T., Lee, Y., Higa, M., Wang, Z.w., Pan, W., Shimomura, I., Unger, R.H.: Leptin, troglitazone, and the expression of sterol regulatory element binding proteins in liver and pancreatic islets. Proc. Natl. Acad. Sci. **97**(15), 8536–8541 (2000)

59. Szklarczyk, D., Morris, J.H., Cook, H., Kuhn, M., Wyder, S., Simonovic, M., Santos, A., Doncheva, N.T., Roth, A., Bork, P., et al.: The string database in 2017: quality-controlled protein–protein association networks, made broadly accessible. Nucleic Acids Res. **45**, D362–D368 (2016). gkw937
60. Kestler, H.A., Müller, A., Gress, T.M., Buchholz, M.: Generalized venn diagrams: a new method of visualizing complex genetic set relations. Bioinformatics **21**(8), 1592–1595 (2004)
61. Martin, B., Chadwick, W., Yi, T., Park, S.S., Lu, D., Ni, B., Gadkaree, S., Farhang, K., Becker, K.G., Maudsley, S.: Vennture–a novel venn diagram investigational tool for multiple pharmacological dataset analysis. PLoS ONE **7**(5), e36,911 (2012)
62. Kestler, H.A., Müller, A., Kraus, J.M., Buchholz, M., Gress, T.M., Liu, H., Kane, D.W., Zeeberg, B.R., Weinstein, J.N.: Vennmaster: area-proportional euler diagrams for functional go analysis of microarrays. BMC Bioinformatics **9**(1), 67 (2008)
63. Chen, H., Boutros, P.C.: Venndiagram: a package for the generation of highly-customizable venn and euler diagrams in r. BMC Bioinformatics **12**(1), 35 (2011)
64. Heberle, H., Meirelles, G.V., da Silva, F.R., Telles, G.P., Minghim, R.: Interactivenn: a web-based tool for theanalysis of sets through venn diagrams. In: Embrapa Informática Agropecuária-Artigo em anais de congresso (ALICE). BMC Bioinformatics, v. 16, p. 1–7 (2015)
65. Hulsen, T., de Vlieg, J., Alkema, W.: Biovenn–a web application for the comparison and visualization of biological lists using area-proportional venn diagrams. BMC Genomics **9**(1), 488 (2008)
66. Wang, Y., Thilmony, R., Gu, Y.Q.: Netvenn: an integrated network analysis web platform for gene lists. Nucleic Acids Res. **42**(W1), W161–W166 (2014)
67. Jeggari, A., Alekseenko, Z., Petrov, I., Dias, J.M., Ericson, J., Alexeyenko, A.: Evinet: a web platform for network enrichment analysis with flexible definition of gene sets. Nucleic Acids Res. **46**(W1), W163–W170 (2018)
68. Hur, B., Kang, D., Lee, S., Moon, J.H., Lee, G., Kim, S.: Venn-dianet: venn diagram based network propagation analysis framework for comparing multiple biological experiments. BMC Bioinformatics **20**(23), 1–12 (2019)
69. Li, Y., Patra, J.C.: Genome-wide inferring gene–phenotype relationship by walking on the heterogeneous network. Bioinformatics **26**(9), 1219–1224 (2010)
70. Chen, J., Aronow, B.J., Jegga, A.G.: Disease candidate gene identification and prioritization using protein interaction networks. BMC Bioinformatics **10**(1), 73 (2009)
71. Chen, J.Y., Shen, C., Sivachenko, A.Y.: Mining alzheimer disease relevant proteins from integrated protein interactome data. In: Biocomputing 2006, pp. 367–378. World Scientific (2006)
72. Szklarczyk, D., Franceschini, A., Wyder, S., Forslund, K., Heller, D., Huerta-Cepas, J., Simonovic, M., Roth, A., Santos, A., Tsafou, K.P., et al.: String v10: protein–protein interaction networks, integrated over the tree of life. Nucleic Acids Res. **43**(D1), D447–D452 (2014)
73. Dirmeier, S.: diffusr: Network Diffusion Algorithms (2018). https://CRAN.R-project.org/package=diffusr. R package version 0.1.4
74. Edgar, R., Domrachev, M., Lash, A.E.: Gene expression omnibus: Ncbi gene expression and hybridization array data repository. Nucleic Acids Res. **30**(1), 207–210 (2002)
75. Spurgeon, M.E., den Boon, J.A., Horswill, M., Barthakur, S., Forouzan, O., Rader, J.S., Beebe, D.J., Roopra, A., Ahlquist, P., Lambert, P.F.: Human papillomavirus oncogenes reprogram the cervical cancer microenvironment independently of and synergistically with estrogen. Proc. Natl. Acad. Sci. **114**(43), E9076–E9085 (2017)
76. NIH: Biological pathways fact sheet (2015). https://www.genome.gov/about-genomics/fact-sheets/Biological-Pathways-Fact-Sheet
77. Kanehisa, M., Goto, S.: Kegg: kyoto encyclopedia of genes and genomes. Nucleic Acids Res. **28**(1), 27–30 (2000)
78. Moon, J.H., Lim, S., Jo, K., Lee, S., Seo, S., Kim, S.: Pintnet: construction of condition-specific pathway interaction network by computing shortest paths on weighted ppi. BMC Syst. Biol. **11**(2), 15 (2017)

79. Lee, S., Lim, S., Lee, T., Sung, I., Kim, S.: Cancer subtype classification and modeling by pathway attention and propagation. Bioinformatics **36**(12), 3818–3824 (2020)
80. Bray, F., Ferlay, J., Soerjomataram, I., Siegel, R.L., Torre, L.A., Jemal, A.: Global cancer statistics 2018: Globocan estimates of incidence and mortality worldwide for 36 cancers in 185 countries. CA Cancer J. Clin. **68**(6), 394–424 (2018)
81. Fisher, B., Costantino, J., Redmond, C., Poisson, R., Bowman, D., Couture, J., Dimitrov, N.V., Wolmark, N., Wickerham, D.L., Fisher, E.R., et al.: A randomized clinical trial evaluating tamoxifen in the treatment of patients with node-negative breast cancer who have estrogen-receptor–positive tumors. New Engl. J. Med. **320**(8), 479–484 (1989)
82. Cho, H.S., Mason, K., Ramyar, K.X., Stanley, A.M., Gabelli, S.B., Denney Jr, D.W., Leahy, D.J.: Structure of the extracellular region of her2 alone and in complex with the herceptin fab. Nature **421**(6924), 756 (2003)
83. Dai, X., Li, T., Bai, Z., Yang, Y., Liu, X., Zhan, J., Shi, B.: Breast cancer intrinsic subtype classification, clinical use and future trends. Am. J. Cancer Res. **5**(10), 2929 (2015)
84. Lee, S., Park, Y., Kim, S.: Midas: Mining differentially activated subpaths of kegg pathways from multi-class rna-seq data. Methods **124**, 13–24 (2017)
85. Huang, M.W., Chen, C.W., Lin, W.C., Ke, S.W., Tsai, C.F.: Svm and svm ensembles in breast cancer prediction. PLoS ONE **12**(1), e0161,501 (2017)
86. Rhee, S., Seo, S., Kim, S.: Hybrid approach of relation network and localized graph convolutional filtering for breast cancer subtype classification. Preprint (2017). arXiv:1711.05859
87. Burt, J.R., Torosdagli, N., Khosravan, N., RaviPrakash, H., Mortazi, A., Tissavirasingham, F., Hussein, S., Bagci, U.: Deep learning beyond cats and dogs: recent advances in diagnosing breast cancer with deep neural networks. Br. J. Radiol. **91**(1089), 20170,545 (2018)
88. Bruna, J., Zaremba, W., Szlam, A., LeCun, Y.: Spectral networks and locally connected networks on graphs. Preprint (2013). arXiv:1312.6203
89. Defferrard, M., Bresson, X., Vandergheynst, P.: Convolutional neural networks on graphs with fast localized spectral filtering. In: Advances in Neural Information Processing Systems, pp. 3844–3852 (2016)
90. Pyszniak, M., Tabarkiewicz, J., Łuszczki, J.J.: Endocannabinoid system as a regulator of tumor cell malignancy–biological pathways and clinical significance. Onco Targets Ther. **9**, 4323 (2016)
91. Jo, K., Jung, I., Moon, J.H., Kim, S.: Influence maximization in time bounded network identifies transcription factors regulating perturbed pathways. Bioinformatics **32**(12), i128–i136 (2016)
92. Mazza, A., Klockmeier, K., Wanker, E., Sharan, R.: An integer programming framework for inferring disease complexes from network data. Bioinformatics **32**(12), i271–i277 (2016)

Printed in the United States
by Baker & Taylor Publisher Services